DECORATION
by GARUHARU

GARUHARU MASTER BOOK SERIES 3

DECORATION by GARUHARU

초판 1쇄 발행	2020년 8월 20일
초판 4쇄 발행	2023년 12월 18일

지은이	윤은영
어시스턴트	김은솔
영문번역	김예성
펴낸이	한준희
발행처	(주)아이콕스

기획·편집	박윤선
디자인	김보라
사진	박성영
스타일링	이화영
영업·마케팅	김남권, 조용훈, 문성빈
경영지원	김효선, 이정민

주소	경기도 부천시 조마루로385번길 122 삼보테크노타워 2002호
홈페이지	www.icoxpublish.com
쇼핑몰	www.baek2.kr (백두도서쇼핑몰)
인스타그램	@thetable_book
이메일	thetable_book@naver.com
전화	032) 674-5685
팩스	032) 676-5685
등록	2015년 07월 09일 제 386-251002015000034호
ISBN	979-11-6426-126-0

DECORATION
by GARUHARU

데 커 레 이 션   바 이   가 루 하 루

윤은영 지음

더 테 이 블
THE TABLE

PROLOGUE

열다섯. 작은 손으로 빚어낸 밀가루 반죽이 오븐 속에서 근사한 과자로 부풀어 오르는 모습을 지켜보았던 날, 제 마음도 따뜻하고 달콤한 과자처럼 부풀어 올랐습니다. 울퉁불퉁 오트밀 쿠키 한 조각을 맛보고 엄지를 추켜세우며 한껏 웃던 가족들과 친구들의 행복한 표정이 저를 '파티시에'라는 길로 이끌어 주었습니다.

파티시에는 다른 이들에게 달콤한 맛의 휴식과 즐거움을 주는 행복한 직업이지만, 기술을 배우고 능숙해지기까지 긴 시간과 노력이 필요했습니다. 하지만 이 길고 힘든 시간 끝에는 저와 제가 만든 제품들이 함께 성장해 있었습니다.

단지 레시피가 아닌 그 과정에서 겪었던 많은 시행착오와 실패를 경험하며 터득한 포인트와 팁, 도구 활용법은 물론 제조 공정에서의 잦은 실수를 줄여주는 방법들을 이 책에 담았습니다. 또한 해외 마스터 클래스를 진행하며 만난 다양한 문화권의 훌륭한 셰프들과 새로운 식재료들로부터 받은 영감의 결과물을 오롯이 담고자 노력했습니다.

이 책이 누군가에게 새로운 영감을 주는 도구로 사용되길 바랍니다. 이 책의 레시피를 토대로 여러분의 주변에서 구할 수 있는 재료를 활용해 다양한 시도를 해보셨으면 좋겠습니다. 제가 늘 작업대 위에서 기존의 제품과 새로운 재료의 조합을 고민하듯, 여러분의 작업실에서도 이러한 고민과 새로운 시도가 계속되길 희망하며, 그 과정에서 저의 책이 작은 도움이 되었으면 합니다.

끝으로 이 책을 펴내는 데 도움을 주신 많은 분들과 아낌없는 지원을 해주신 더테이블 관계자 분들께 고마운 마음을 전합니다.

윤은영

Fifteen. The day I watched the dough I made with my small hands rise up into gorgeous cookies, my heart billowed just like the warm and sweet cookie. The happy faces of family and friends who tasted a piece of the lumpy oatmeal cookie, raising their thumbs up with big loving laughers, led me to the path of becoming a 'pâtissier'.

Being a pâtissier is blissful job, giving others sweet taste of relaxation and delightfulness, but it took a long time and effort to learn the skills and become proficient. However, at the end of this long and challenging time, I and the products I made have been growing together.

This book is not just a collection of recipes, but also reflects a lot of trials and errors that we have experienced along the way, and tried to cover the points and tips we've learned, including how to utilize tools, as well as ways to minimize frequent mistakes during process of making. I also tried to capture the results of inspirations from the talented chefs we've met from various cultures and new ingredients we came across while conducting the overseas master classes.

I hope this book will serve as a new inspiration for someone. Please make variety of attempts with ingredients available near you by utilizing the recipes in this book. Just as I always contemplate the combination of existing products and new ingredients on my workstation, I hope these thoughts and new attempts will continue in your studio, and wish my book will be of little help along the way.

And last but not least, I would like to thank many people who helped me publish this book, and to the officials of THETABLE publishers for their generous support.

Yun Eunyoung

CONTENTS

PART 06. **Etc**

PART 07. **Nature**

PREPARATION

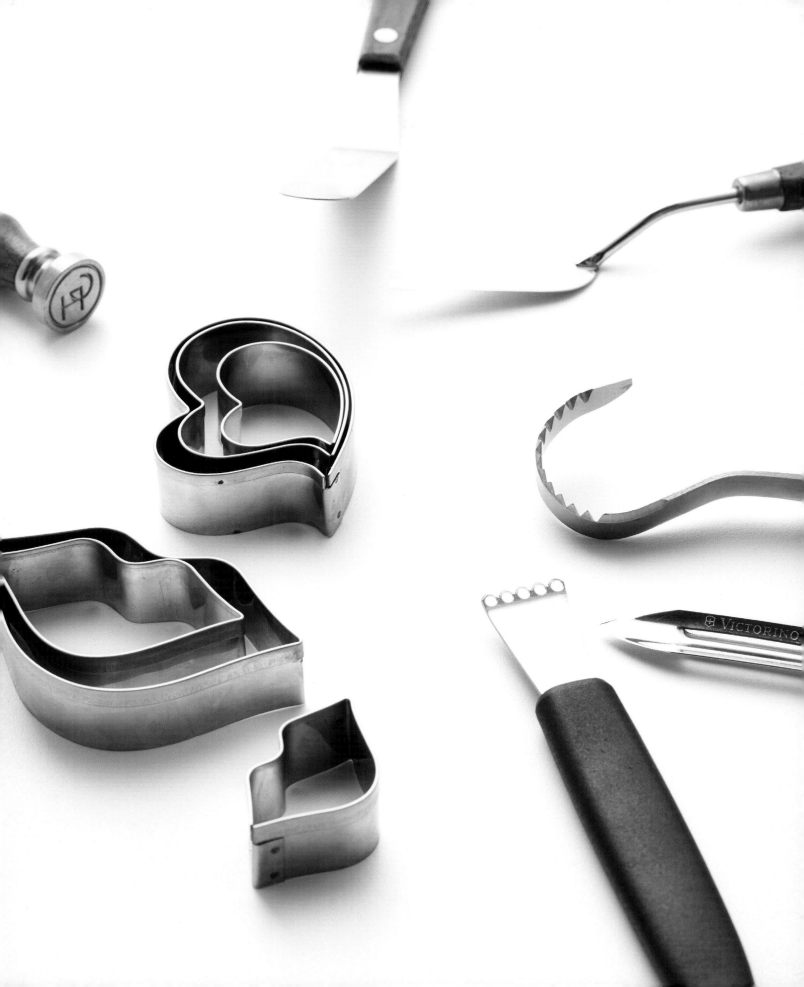

도구 Tools

인장 [STAMP]

스패출러
[SPATULA]

데커레이션 나이프
[DECORATION KNIFE]

필러 [PEELER]

감귤류 필러
[CITRUS ZESTER]

모양 커터
[SHAPE CUTTERS]

스패출러

반죽이나 크림을 펼치거나 고르게 정리할 때 사용합니다. 손잡이와 날이 일직선인 것, L자로 굽은 것 두 가지 종류가 있으며 사이즈가 다양하므로 작업에 맞춰 선택할 수 있습니다.

모양 커터

비스킷 반죽이나 구워진 비스퀴, 초콜릿 장식물 등을 모양내어 커팅하는 데 사용합니다. 다양한 모양과 재질의 커터가 있으며 3D 프린터를 이용해 원하는 모양으로 제작할 수도 있습니다.

인장

동 재질의 인장은 차갑게 하여 초콜릿 데커레이션 용도로 사용할 수 있으며 열을 가해 구운 비스퀴 또는 과자 반죽에 문양을 낼 수도 있습니다.

필러

과일이나 채소의 껍질을 제거하는 데 주로 사용하며, 코코넛 데커레이션 작업 시 코코넛을 얇게 슬라스하는 용도로 사용할 수 있습니다.

감귤류 필러

레몬이나, 라임, 오렌지, 자몽 등 감귤류의 껍질을 모양내어 필링하는 데 사용합니다.

데커레이션 나이프

초콜릿, 잔두야, 버터 등을 모양내는 데 사용합니다.

Spatula

It is usually used to spread or even out dough/batter or cream. There are two types; one that's straight from the handle to the blade, and one in which the blade is bent to L-shape, and various sizes are available, which can be chosen according to needs.

Shape cutters

These are used to cut biscuit dough, baked biscuits, and chocolate decorations in desired shapes. Various shapes and materials are available, and customized shapes can be made with a 3D printer.

Stamp

A stamp made with copper can be used cold to make chocolate decorations or heated to make imprints on the baked goods.

Peeler

Mainly used to remove the peel of fruits or vegetables, it can also be used to thinly slice coconut to use as decoration.

Citrus zester

It is used to zest the peel of lemon, lime, or orange.

Decoration knife

It is used to shape chocolate, gianduja, butter, etc.

진공 성형기
Vacuum Former

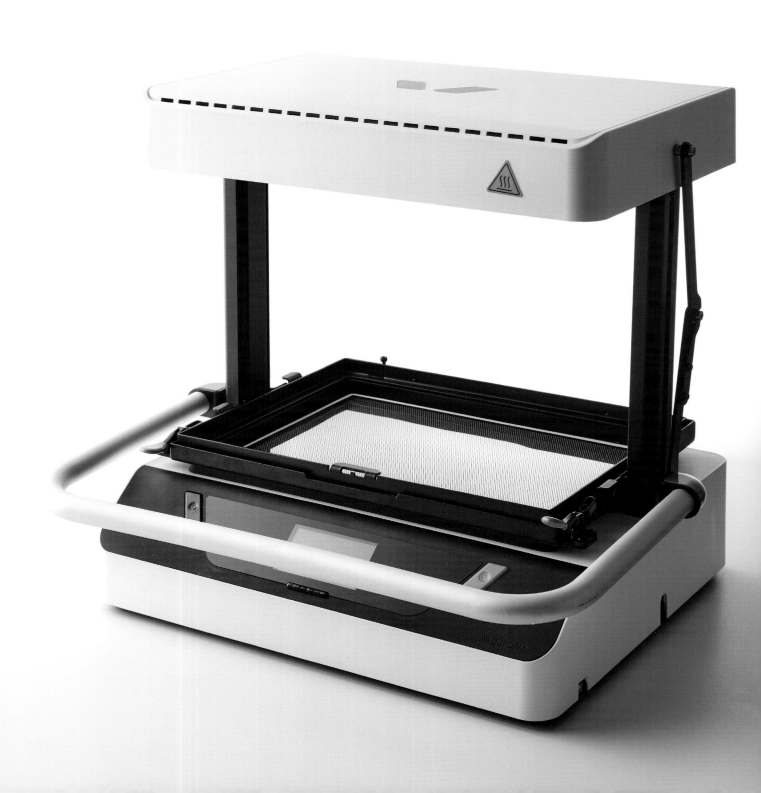

몰드를 제작하는 데 사용하는 도구입니다. 가열된 열 가소성 시트를 진공·흡착시켜 모양을 만드는 원리로 일상생활에서 사용하는 다양한 오브제 또는 점토나 3D 프린터를 이용해 제작한 디자인 오브제를 사용해 커스텀 몰드를 제작할 수 있습니다. 사용 목적에 맞게 시트의 재질, 두께를 결정할 수 있으며 식품용으로 사용할 경우에는 반드시 독성이 없는 안정성이 입증된 시트를 선택해야 합니다.

It is a tool to produce molds. Its principle is to make shapes by vacuum-adhering the heated thermoplastic sheets. Custom mold can be made with different kinds of objects used in everyday life or objects made with clay or 3D printers. The purpose of use can determine the material and thickness. However, when used for food, it is crucial to select sheets proven to be stable and non-toxic.

커팅기
Cutting Machine

디자인한 도안을 커팅해주는 도구입니다. 커팅된 필름을 이용해 초콜릿 장식물 작업이 가능하며, 분사하거나 파우더를 모양내어 도포할 수도 있습니다. 사용 목적에 맞게 재질과 두께를 선택할 수 있습니다.

It is a tool that cuts designed patterns. Chocolate decorations can be made with cut films, and they can also be sprayed on or covered with powders in desired designs. Material and thickness can be determined according to the purpose of use.

몰드 제작용 실리콘

Silicones for Mold Making

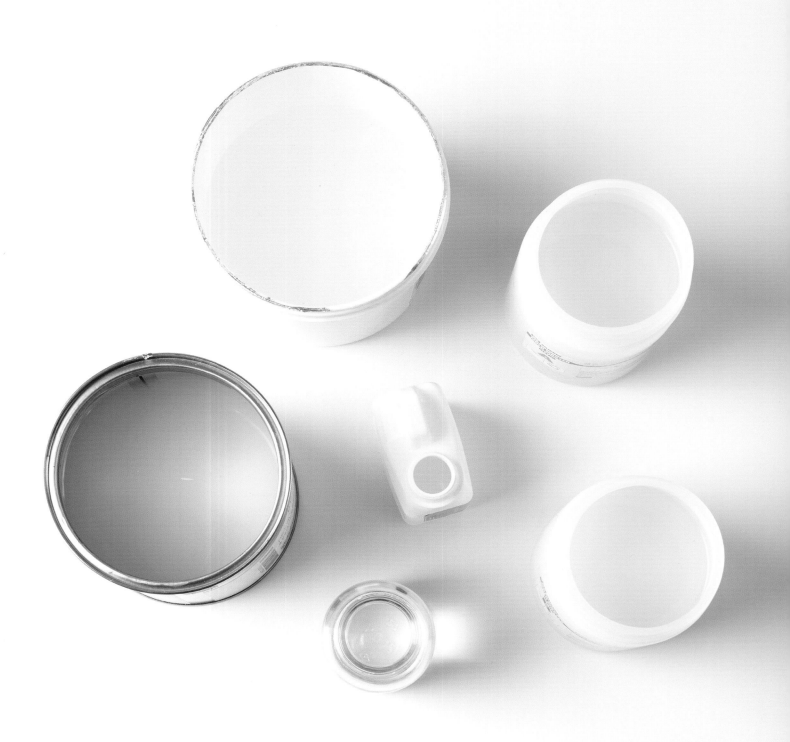

몰드 제작용 실리콘은 축합형과 부가형 두 가지 종류가 있습니다.

식품용 몰드를 제작할 때는 경화 시 부산물 생성이 없고 부식되지 않는 부가형 실리콘이 적합하며, 반드시 독성이 없는 안정성이 입증된 제품을 선택해야 합니다. 또한 동일한 타입의 실리콘이라도 점도와 경도에 따라 다양한 제품들이 있으며 사용 가능한 온도 범위, 색상, 투명도 등의 차이가 있으므로 사용 목적에 맞게 선택할 수 있습니다.

각 제품들은 경화제와의 배합 비율, 작업 시간, 경화 시간의 차이가 있기 때문에 사용 전 공급자에게 정확한 제품 사용 방법을 확인하는 것이 좋습니다.

There are two types of silicones for making molds; Condensation-type and Addition-type.

When making food-grade molds, an Addition-type of silicone, which does not corrode nor create by-products during hardening, is suitable. Make sure it is proven to be safe without any toxic elements.

Also, even if it's the same type of silicone, various products are available depending on the viscosity and hardness and have a difference in the temperature range of usability, color, transparency, etc., which can be selected according to the purpose of use.

Each product has a difference in mixing ratio with the hardening agent, working time, and curing time. Therefore, it is best to check with the supplier for the correct usage.

CHOCOLATE TEMPERING

초콜릿 템퍼링(접종법)

접종법은 녹인 초콜릿에 안정된 결정 상태의 초콜릿을 넣어 템퍼링하는 방법입니다. 여러 가지 템퍼링 방법 중 작업 방식이 간편해 대량 작업 시에도 비교적 손쉽게 템퍼링할 수 있습니다.

Seeding method is a technique of tempering, where stable crystallized chocolate is added to melted chocolate. This method is simple among various methods of tempering, which also makes it relatively easy to work with large amount of chocolate.

ingredients

커버추어 초콜릿

couverture chocolate

Process

1. 커버추어 초콜릿을 폴리카보네이트 볼에 담고 녹여준다.
(다크초콜릿 55~58℃, 밀크초콜릿 45~48℃, 화이트초콜릿 45~48℃)

2. 녹인 초콜릿 양의 30% 정도의 초콜릿을 1에 넣고 골고루 저어준 후 1분간 그대로 둔다.

3. 핸드블렌더를 이용해 초콜릿 덩어리가 남지 않도록 균일하게 믹싱하며 온도를 낮춰준다.
(다크초콜릿 31~32℃, 밀크초콜릿 29~30℃, 화이트초콜릿 28~29℃)
* 이때 빠른 속도로 장시간 믹싱하면 마찰열에 의해 온도가 과도하게 올라갈 수 있으므로 주의한다.

4. 스패출러 또는 나이프 끝부분에 템퍼링 테스트를 한 후 사용한다.
* 23~24℃를 유지한 작업실에서 5분 이내에 얼룩 없이 매끄럽게 초콜릿이 굳으면 템퍼링이 잘된 상태이다.

1. Melt couverture chocolate in a polycarbonate bowl. (Dark chocolate 55~58℃, milk chocolate 45~48℃, white chocolate 45~48℃)

2. Stir in evenly about 30% of the melted chocolate to 1, and let stand for 1 minute.

3. Reduce the temperature of chocolate using hand blender, so that it's mixed evenly and no chocolate chunks are left.
(Dark chocolate 31~32℃, milk chocolate 29~30℃, white chocolate 28~29℃)
* Keep in mind when mixing at a high speed for a long time, temperature can excessively increase due to heat caused by friction.

4. Test on the tip of a spatula or knife before use.
* If chocolate hardens smoothly within 5 minutes in a studio that maintained 23~24℃, the tempering is done properly.

CORNET

코르네

삼각형 모양으로 준비한 베이킹 페이퍼를 고깔 모양으로 말아준 후 끝부분을 안쪽으로 접어 풀리지 않
도록 고정시켜 사용합니다.

Prepare baking paper cut into triangle, and roll into cone shape. Fold the end
inward to fasten, so it won't undo itself.

PART 01

CHOCOLATE

초콜릿의 특성을 이용해 다양한 표현이 가능합니다. 모양 커터나 도안을 활용해 단순한 모양의 장식물 작업이 가능하며, 몰드나 파이프 등을 활용해 입체감 있는 장식물을 만들 수도 있습니다.

It is possible to make various expressions using the characteristic of chocolate. Simple shapes of decorations can be made using shape cutters or pre-made designs and make three-dimensional decorations with molds or pipes.

CHOCOLATE SHEET

초콜릿 시트

tools & ingredients

| TOOLS | 투명 필름, 밀대, 체 | clear plastic films, rolling pin, sieve |
| INGREDIENTS | 커버추어 초콜릿, 카카오 파우더 | couverture chocolate, cacao powder |

Process

1. 두 장의 투명 필름 사이에 템퍼링한 커버추어 초콜릿을 적당량 부어준다.

   * 초콜릿을 붓기 전 카카오파우더, 녹차파우더, 과일파우더 등을 이용해 필름을 장식할 수 있다.

2. 필름을 덮고 밀대를 이용해 균일한 두께로 밀어 편다.

3. 완전히 굳으면 자연스럽게 부서진 모양을 살려 사용한다.

1. Pour a moderate amount of tempered couverture chocolate between two sheets of clear plastic film.

   * Before pouring the chocolate, the film can be decorated with cacao powder, green tea powder, fruit powders, etc.

2. Cover with the film, and roll into even thickness using a rolling pin.

3. After it's completely crystallized, use the naturally broken shapes.

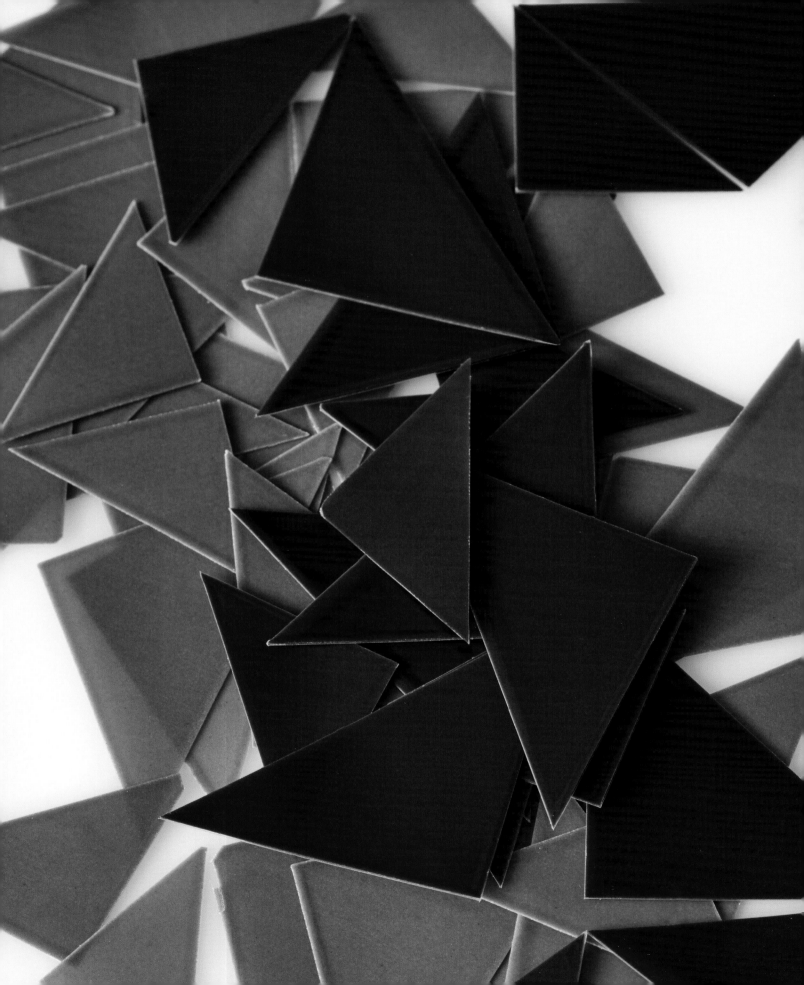

KNIFE CUTTING

나이프 커팅

tools & ingredients

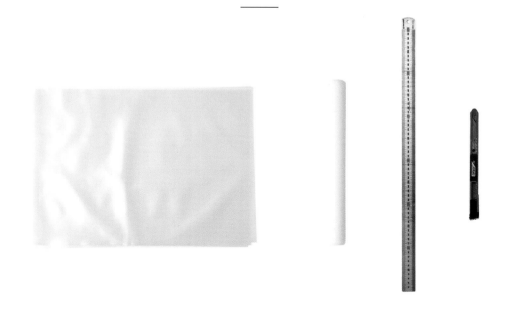

TOOLS	투명 필름, 밀대, 자, 칼		clear plastic films, rolling pin, ruler, exacto knife
INGREDIENTS	커버추어 초콜릿		couverture chocolate

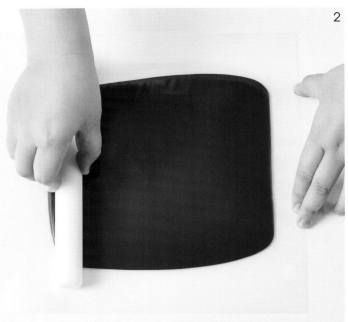

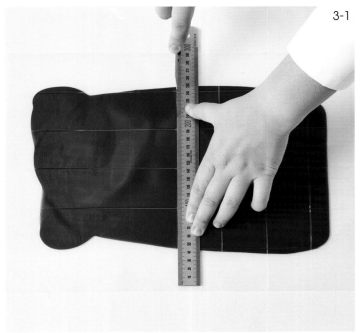

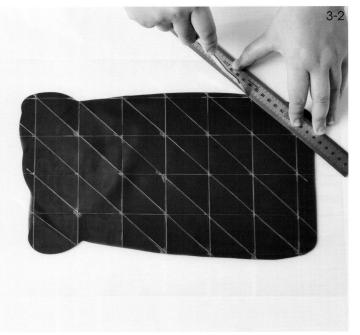

Chocolate

Process

1. 두 장의 투명 필름 사이에 템퍼링한 커버추어 초콜릿을 적당량 부어준다.

2. 필름을 덮고 밀대를 이용해 균일한 두께로 밀어 편다.

3. 나이프를 이용해 원하는 모양으로 커팅한다.

   * 원하는 사이즈와 모양으로 커팅할 수 있으며 도안을 이용해 커팅할 수도 있다.

1. Pour a moderate amount of tempered couverture chocolate between two sheets of clear plastic film.

2. Cover with the film, and roll into even thickness using a rolling pin.

3. Cut into desired shapes using a knife.

   * It can be cut into desired sizes and shapes, or a pre-made design can be used as a guide.

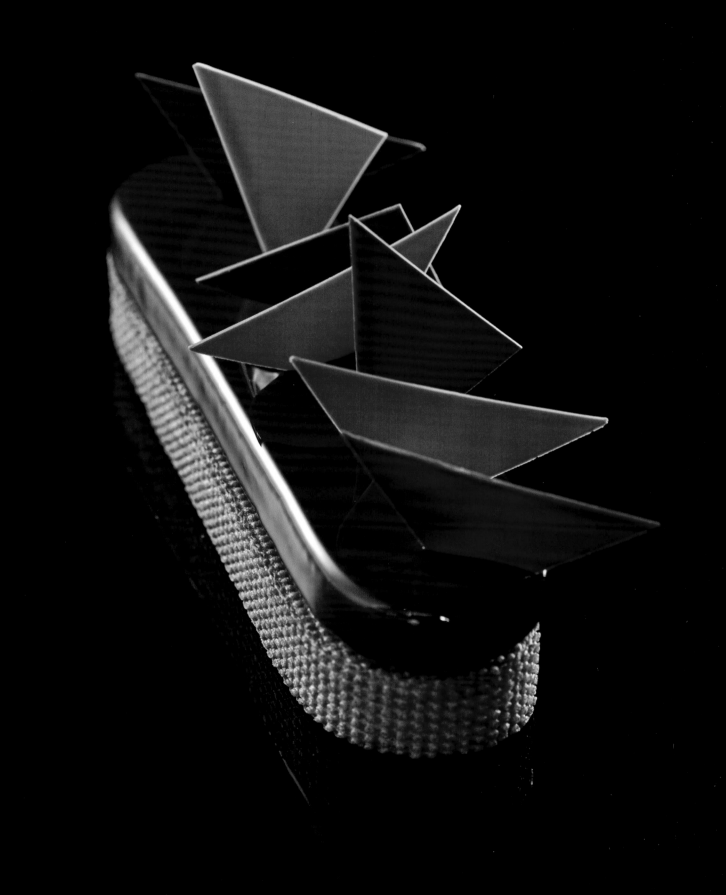

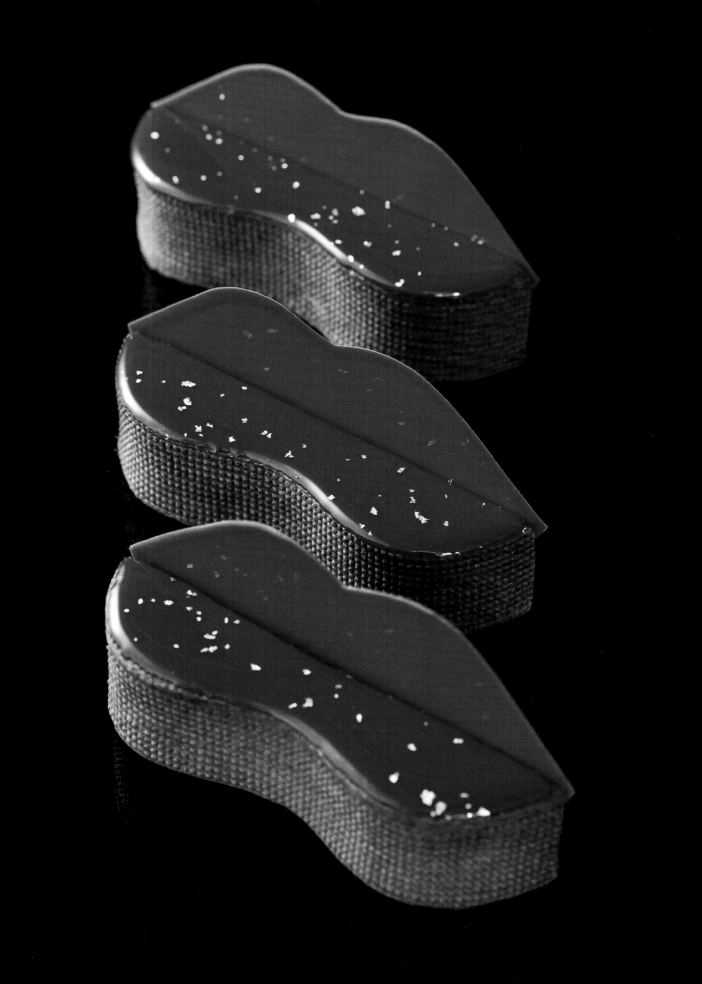

CUTTER CUTTING

커터 커팅

tools & ingredients

TOOLS	투명 필름, 밀대, 모양 커터	clear plastic films, rolling pin, shape cutters
INGREDIENTS	커버추어 초콜릿	couverture chocolate

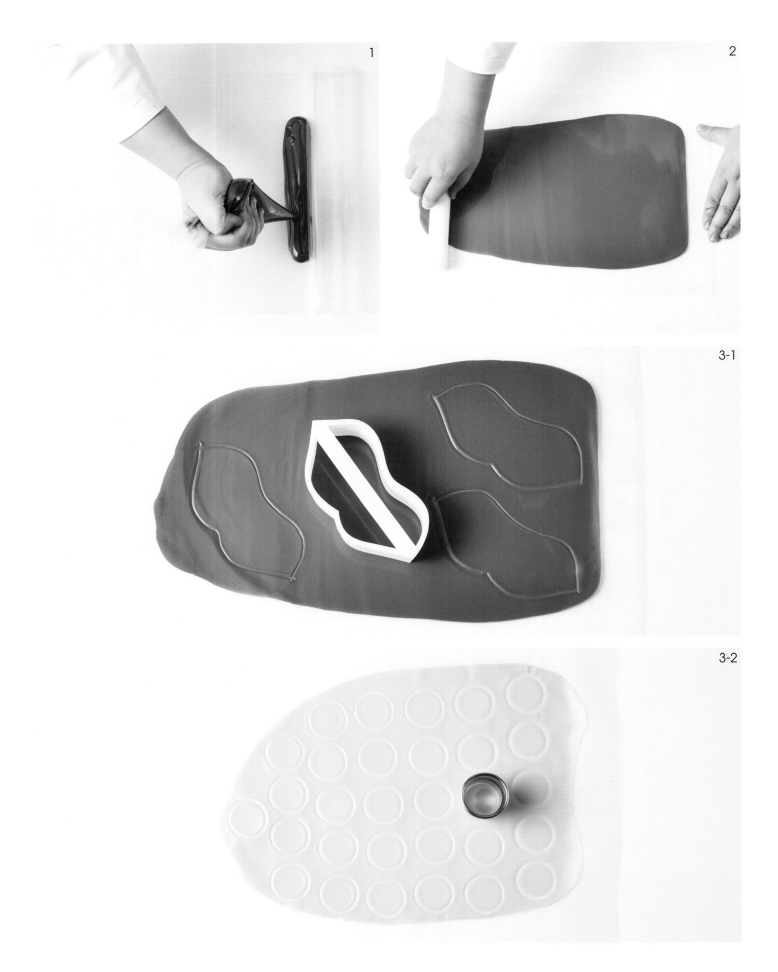

1

2

3-1

3-2

Process

1. 두 장의 투명 필름 사이에 템퍼링한 커버추어 초콜릿을 적당량 부어준다.

2. 필름을 덮고 밀대를 이용해 균일한 두께가 되도록 밀어 편다.

3. 원하는 모양의 커터를 이용해 커팅한다.

1. Pour a moderate amount of tempered couverture chocolate between two sheets of clear plastic film.

2. Cover with the film, and roll into even thickness using a rolling pin.

3. Cut by using the desired shape of the cutter.

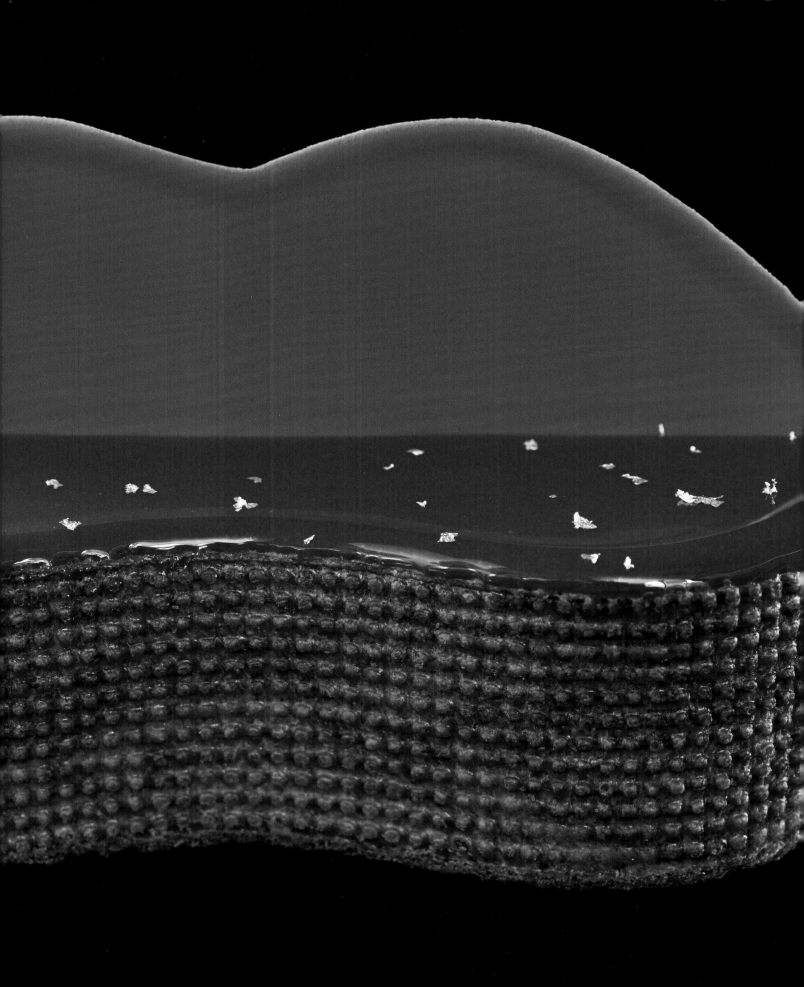

CHOCOLATE TECHNIQUE 04

SCRATCHING

스크래치

tools & ingredients

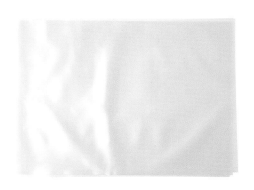 

| TOOLS | 투명 필름, 밀대, 거친 솔 | clear plastic films, rolling pin, stiff brush |
| INGREDIENTS | 커버추어 초콜릿 | couverture chocolate |

Process

1. 두 장의 투명 필름 사이에 템퍼링한 커버추어 초콜릿을 적당량 부어준다.

2. 필름을 덮고 밀대를 이용해 균일한 두께가 되도록 밀어 편다. 초콜릿이 충분히 수축하면 필름을 제거한다.

   * 초콜릿이 굳기 전 나이프나 커터를 이용해 원하는 모양으로 커팅할 수 있다.

3. 거친 솔로 긁어 스크래치를 낸다.

1. Pour a moderate amount of tempered couverture chocolate between two sheets of clear plastic film.

2. Cover with the film, and roll into even thickness with a rolling pin. Remove the film when the chocolate is crystallized enough.

   * It can be cut to desired shapes with a knife or cutters before the chocolate sets.

3. Scratch the surface using a stiff brush.

CHOCOLATE TECHNIQUE 05
CURVE-1

굴곡-1

tools & ingredients

TOOLS	투명 필름, 밀대, 아크릴 파이프, 원형 커터	clear plastic films, rolling pin, acrylic pipe, round cutter
INGREDIENTS	커버추어 초콜릿	couverture chocolate

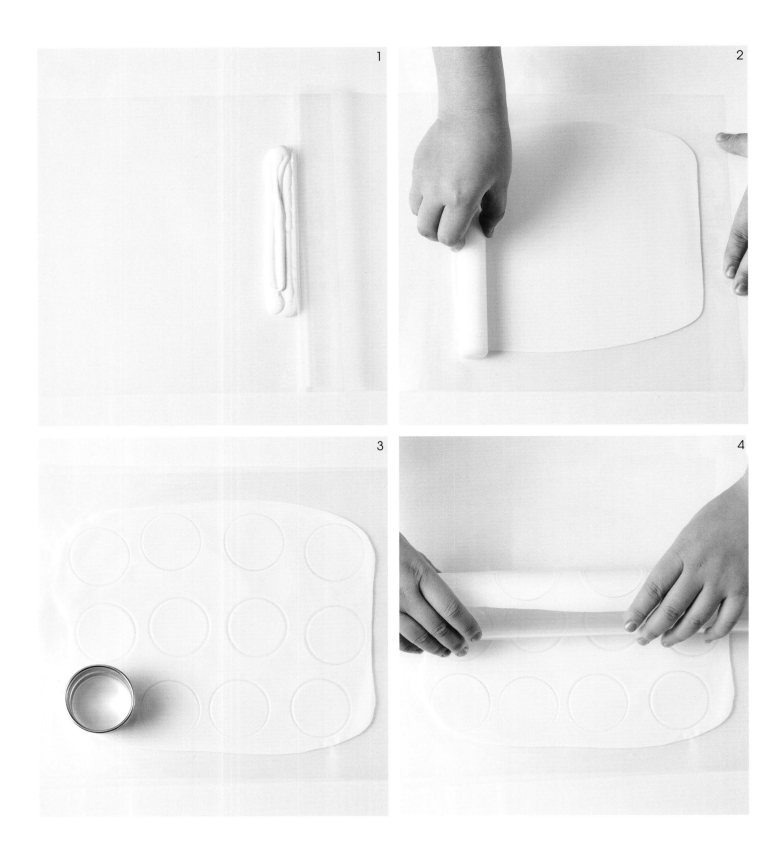

Chocolate

Process

1. 두 장의 투명 필름 사이에 템퍼링한 커버추어 초콜릿을 적당량 부어준다.

2. 필름을 덮고 밀대를 이용해 균일한 두께가 되도록 밀어 편다.

3. 원하는 모양으로 커팅한다.

4. 아크릴 파이프에 감아준다.

   * 아크릴 파이프의 사이즈에 따라 굴곡의 정도를 조절할 수 있습니다.

1. Pour a moderate amount of tempered couverture chocolate between two sheets of clear plastic film.

2. Cover with the film, and roll into even thickness with a rolling pin.

3. Cut into desired shapes.

4. Wrap around an acrylic pipe.

   * Degree of curves can be adjusted according to the size of the acrylic pipes.

CURVE-2

굴곡-2

tools & ingredients

TOOLS	투명 필름, 밀대, 아크릴 파이프	clear plastic films, rolling pin, acrylic pipe
INGREDIENTS	커버추어 초콜릿	couverture chocolate

Process

1. 두 장의 투명 필름 사이에 그린티파우더를 섞어 템퍼링한 화이트초콜릿 적당량을 부어준다.

2. 필름을 덮고 밀대를 이용해 균일한 두께가 되도록 밀어편다.

3. 아크릴 파이프에 감아준다.

4. 충분히 세팅시킨 후 그대로 풀어준다.

5. 자연스럽게 깨진 모양을 살려 필름에서 떼어낸 다음 냉동고에 30분간 넣어둔다.

6. 화이트초콜릿과 카카오버터를 1:1 비율로 섞고 그린티파우더를 첨가해 원하는 색으로 맞춘다.

7. 핸드블렌더로 혼합한다.

8. 차가워진 초콜릿 장식물 표면에 분사해 벨벳 같은 느낌을 표현한 후 현미녹차를 적당량 뿌려 마무리한다.

1. Temper white chocolate mixed with green tea powder, and pour moderate amount between two sheets of clear plastic film.

2. Cover with the film, and roll into even thickness with a rolling pin.

3. Wrap around an acrylic pipe.

4. Set sufficiently and unwrap as is.

5. Remove the film, saving the naturally broken shapes. Keep in the freezer for 30 minutes.

6. Mix white chocolate and cacao butter in a 1:1 ratio, and mix green tea powder to adjust to the desired color.

7. Mix using a hand blender.

8. Spray on the surface of the chilled chocolate decoration to express velvet-like texture, and sprinkle a moderate amount of brown rice green tea to finish.

TOPPING

토핑

tools & ingredients

TOOLS	베이킹 페이퍼, 스패출러, 체, 모양 커터	baking paper, spatula, sieve, shape cutter
INGREDIENTS	커버추어 초콜릿, 코코넛파우더	couverture chocolate, coconut powder

Chocolate

Process

1.   베이킹 페이퍼에 템퍼링한 커버추어 초콜릿을 적당량 부어준다.

2.   스패츌러를 이용해 균일하게 밀어 편다.

3.   초콜릿이 굳기 전에 체를 이용해 코코넛파우더를 토핑한다.
    * 코코넛파우더 외에 다양한 토핑물을 사용할 수 있다.

4.   원하는 모양의 커터로 커팅한다.

1.   Pour a moderate amount of tempered couverture chocolate on a baking paper.

2.   Spread evenly with a spatula.

3.   Sprinkle coconut powder using a sieve before the chocolate sets.
    * Various toppings can be used in addition to coconut powder.

4.   Cut with desired cutter.

CHOCOLATE TECHNIQUE 08
PIPING-1

파이핑-1

tools & ingredients

TOOLS	베이킹 페이퍼, 붓, 체	baking paper, brush, sieve
INGREDIENTS	커버추어 초콜릿, 카카오 파우더	couverture chocolate, cacao powder

Chocolate

Process

1. 종이에 원하는 모양으로 가이드 라인을 그린다.

2. 가이드 종이 위에 베이킹 페이퍼를 올린다. 템퍼링한 커버추어 초콜릿을 불규칙한 모양으로 파이핑한다.

3. 초콜릿이 완전히 굳기 전에 카카오파우더를 도포한다.

4. 초콜릿이 완전히 굳으면 붓으로 여분의 카카오파우더를 털어낸다.

1. On a paper, draw guide lines in desired shapes.

2. Place a sheet of baking paper over the guide paper. Pipe tempered couverture chocolate in irregular shapes.

3. Cover with cacao powder before the chocolate sets completely.

4. When the chocolate is completely crystallized, brush off the excess powder.

CHOCOLATE TECHNIQUE 09
PIPING-2

파이핑-2

tools & ingredients

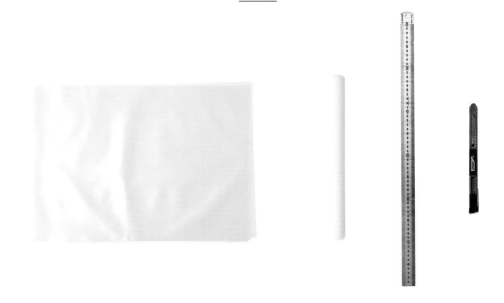

TOOLS	투명 필름, 밀대, 자, 칼	clear plastic films, rolling pin, ruler, exacto knife
INGREDIENTS	커버추어 초콜릿	couverture chocolate

Chocolate

Process

1. 두 장의 투명 필름 사이에 템퍼링한 커버추어 초콜릿을 적당량 부어준다.

2. 필름을 덮고 밀대를 이용해 균일한 두께가 되도록 밀어 편다.

3. 원하는 모양으로 커팅한다. 초콜릿이 충분이 수축하면 필름을 제거한다.

4. 템퍼링한 커버추어 초콜릿을 불규칙한 모양으로 파이핑한다.

1. Pour a moderate amount of tempered couverture chocolate between two sheets of clear plastic film.

2. Cover with the film, and roll into even thickness using a rolling pin.

3. Cut into desired shapes. When the chocolate is crystallized enough, remove the film.

4. Pipe tempered couverture chocolate in irregular shapes.

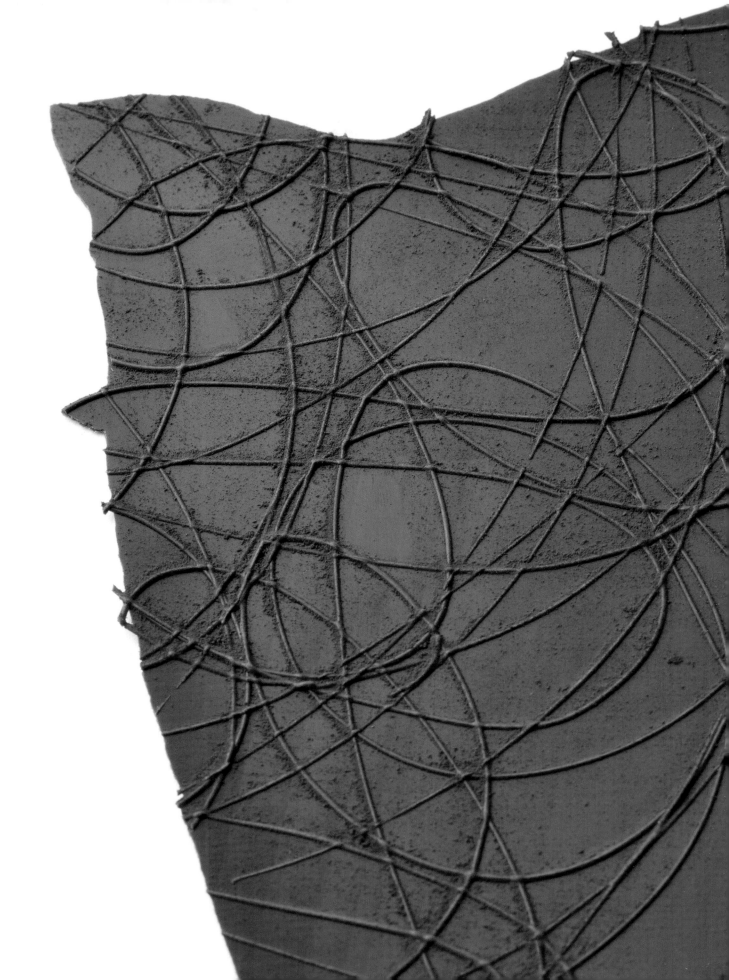

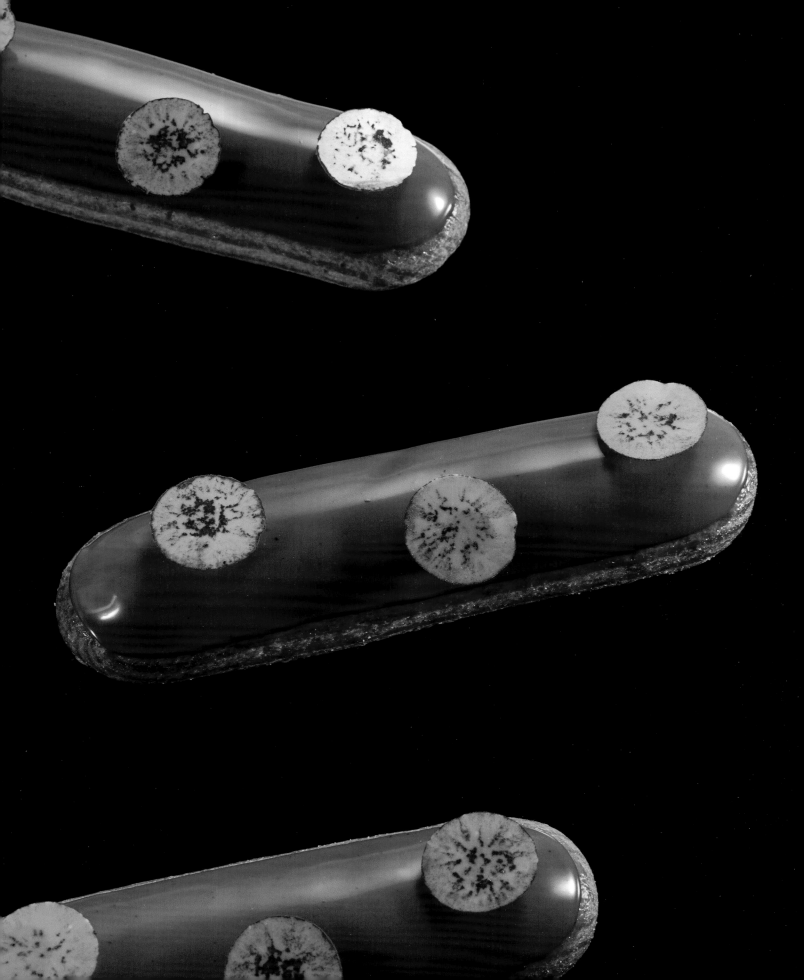

PIPING-3

파이핑-3

tools & ingredients

TOOLS	투명 필름, 체, 누름 도구	clear plastic films, sieve, pressing tool
INGREDIENTS	커버추어 초콜릿, 카카오 파우더	couverture chocolate, cacao powder

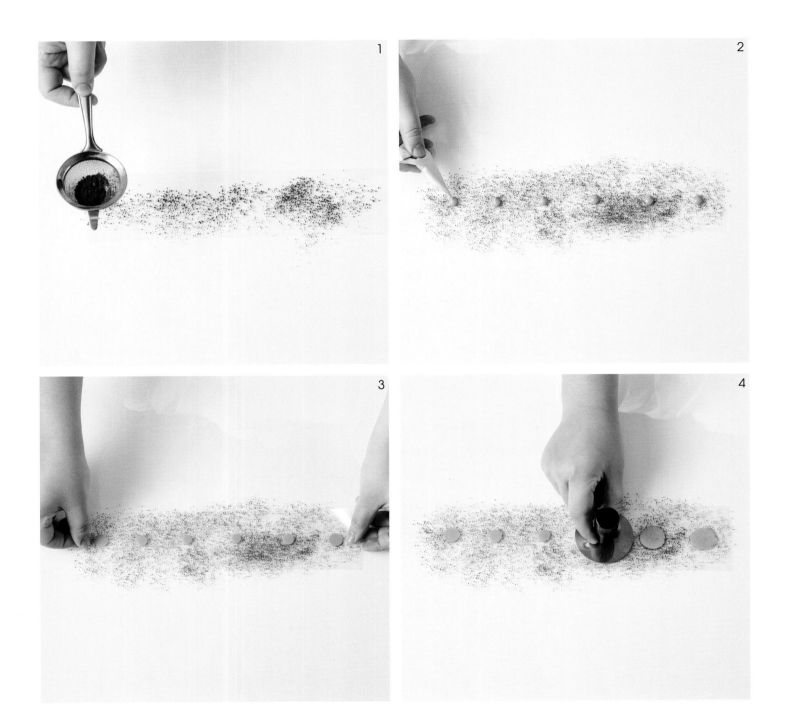

Process

1.   초콜릿 필름 위에 카카오파우더를 적당량 뿌려준다.

2.   템퍼링한 커버추어 초콜릿을 파이핑한다.

3.   파이핑한 초콜릿 위에 다시 필름을 덮어준다.

4.   누름 도구를 이용해 천천히 눌러 초콜릿이 자연스럽게 퍼지게 한다.

1.   Sprinkle a moderate amount of cacao powder on a plastic film.

2.   Pipe tempered couverture chocolate.

3.   Cover with another sheet of plastic film over the piped chocolates.

4.   Press slowly with a pressing tool to let the chocolate spread out naturally.

CHOCOLATE TECHNIQUE 11

TEXTURED SHEET

텍스처 시트

tools & ingredients

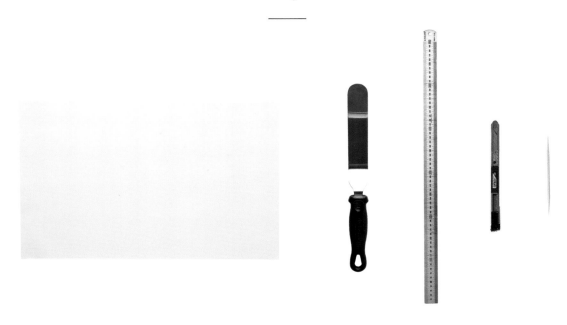

TOOLS	텍스처 시트, 스패출러, 자, 칼, 뾰족한 도구	textured sheet, spatula, ruler, exacto knife, pointed tool
INGREDIENTS	커버추어 초콜릿	couverture chocolate

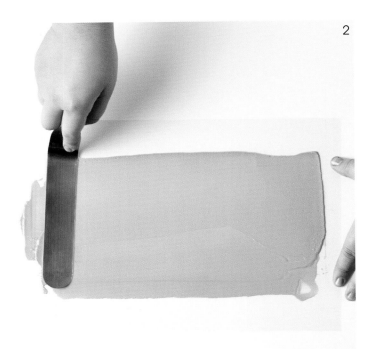

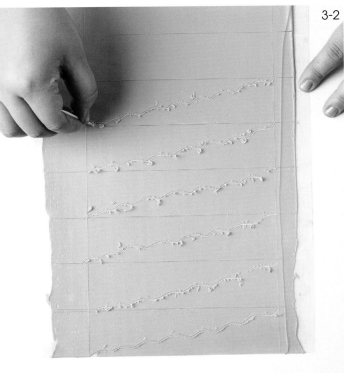

Chocolate

Process

1. 텍스처 시트 위에 템퍼링한 커버추어 초콜릿을 적당량 부어준다.

2. 스패츌러를 이용해 균일하게 밀어 편다.

3. 원하는 모양으로 커팅한다. 충분히 수축시킨 후 텍스처 시트를 제거한다.

1. Pour a moderate amount of tempered couverture chocolate over the textured sheet.

2. Spread evenly with a spatula.

3. Cut into desired shapes. When the chocolate is crystallized enough, remove the textured sheet.

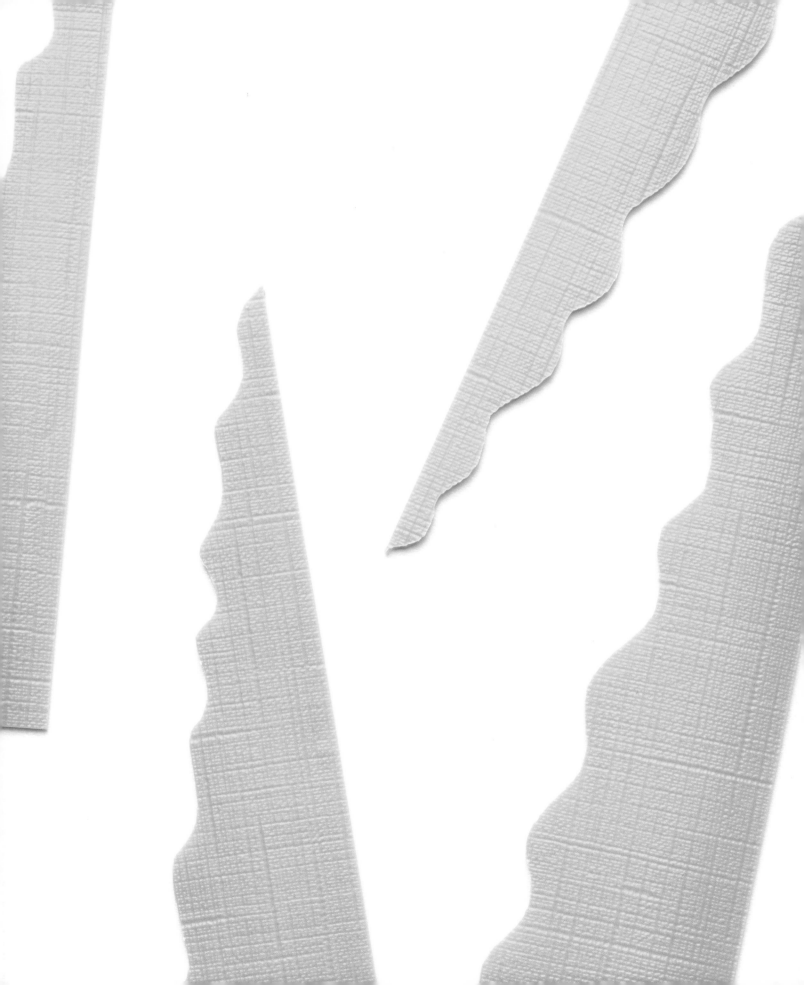

STAMP

스탬프

tools & ingredients

TOOLS	투명 필름, 인장	clear plastic films, stamp
INGREDIENTS	커버추어 초콜릿	couverture chocolate

1

2

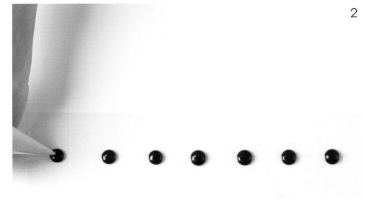

3

1

2

3

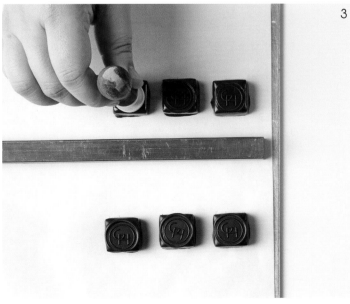

Process

성형 ❶

1. 사용할 인장은 미리 냉동고에 넣어두어 차가운 상태로 준비한다.

2. 투명 필름 위에 템퍼링한 커버추어 초콜릿을 적당량 파이핑한다.

3. 차가운 상태의 인장으로 파이핑한 초콜릿을 눌러 문양을 만든다.

성형 ❷

1. 사용할 인장은 미리 냉동고에 넣어두어 차가운 상태로 준비한다.

2. 가나슈를 디핑한다.

3. 커버추어 초콜릿이 완전히 굳기 전에 차가운 상태의 인장으로 눌러 문양을 만든다.

FORMATION ❶

1. Place the stamp in the freezer in advance to keep it cold until use.

2. On a sheet of clear plastic film, pipe a moderate amount of tempered couverture chocolate.

3. Press the piped chocolate with the cold stamp to make imprints.

FORMATION ❷

1. Place the stamp in the freezer in advance to keep it cold until use.

2. Dip the ganache in chocolate.

3. Press with the cold stamp to make imprints before the chocolate completely crystallizes.

FEATHER

깃털

tools & ingredients

TOOLS	투명 필름, 나이프	clear plastic films, paring knife
INGREDIENTS	커버추어 초콜릿	couverture chocolate

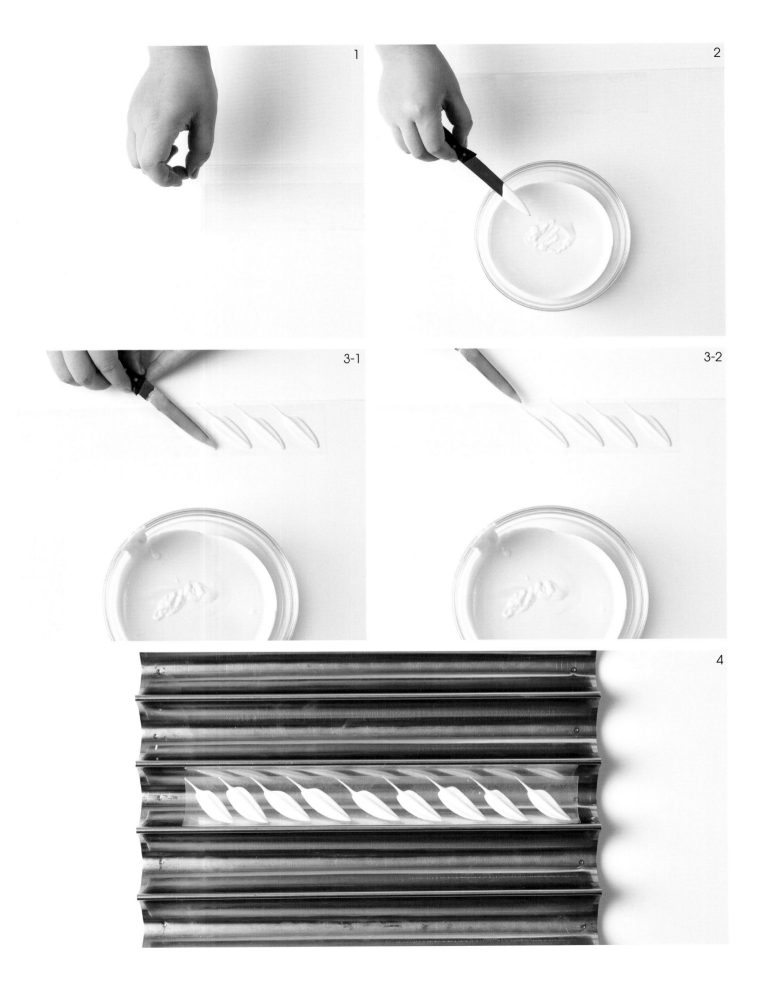

Process

1. 테이블 위에 투명 필름을 고정시킨다.

2. 템퍼링한 커버추어 초콜릿에 나이프를 디핑한다.

3. 나이프를 필름 위에 가볍게 눌러 찍어 깃털 문양을 만든다.

4. 초콜릿이 완전히 굳기 전에 굴곡이 있는 팬 위에 올려 자연스러운 깃털 모양을 만든다.

   * 따뜻하게 가열한 나이프를 이용해 깃털 결을 표현할 수 있다.

1. Fix a clear plastic film on a table.

2. Dip the knife in the tempered couverture chocolate.

3. Lightly stamp the knife on the film to make feather shapes.

4. Place the film on a curved surface before the chocolate sets to make the natural shapes of feathers.

   * Feather vanes can be made with a warm knife.

CHOCOLATE BAND

초콜릿 띠

tools & ingredients

TOOLS	투명 필름, 스패출러, 칼, 아크릴 파이프	clear plastic films, spatula, exacto knife, acrylic pipe
INGREDIENTS	커버추어 초콜릿	couverture chocolate

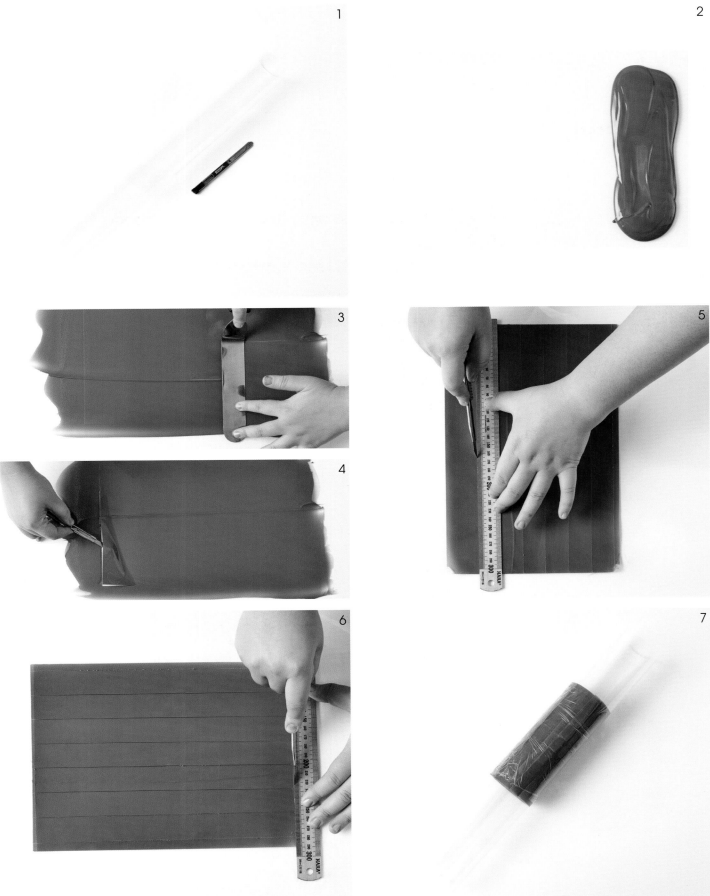

Process

1. 원하는 사이즈 원형 틀 또는 파이프를 준비한다. 틀 또는 파이프 둘레에 맞춰 투명 필름을 재단한다.

2. 재단한 필름 위에 템퍼링한 커버추어 초콜릿을 적당량 부어준다.

3. 스패츌러를 이용해 초콜릿을 얇게 밀어 편다.

4. 초콜릿이 굳기 시작하면 조심스럽게 필름을 들어올린다.

5. 원하는 높이로 커팅한다.

6. 가장자리를 정리한다.

7. 초콜릿이 완전히 굳기 전에 준비한 틀 또는 파이프에 감아준다.

1. Prepare round ring mold or pipe of preferred size. Cut the clear plastic film according to the circumference of the mold or pipe.

2. Pour a moderate amount of tempered couverture chocolate on the cut film.

3. Spread thinly with a spatula.

4. When the chocolate starts to crystallize, carefully lift the film.

5. Cut to the desired height.

6. Organize the ends.

7. Wrap around the prepared mold or pipe before the chocolate completely crystallizes.

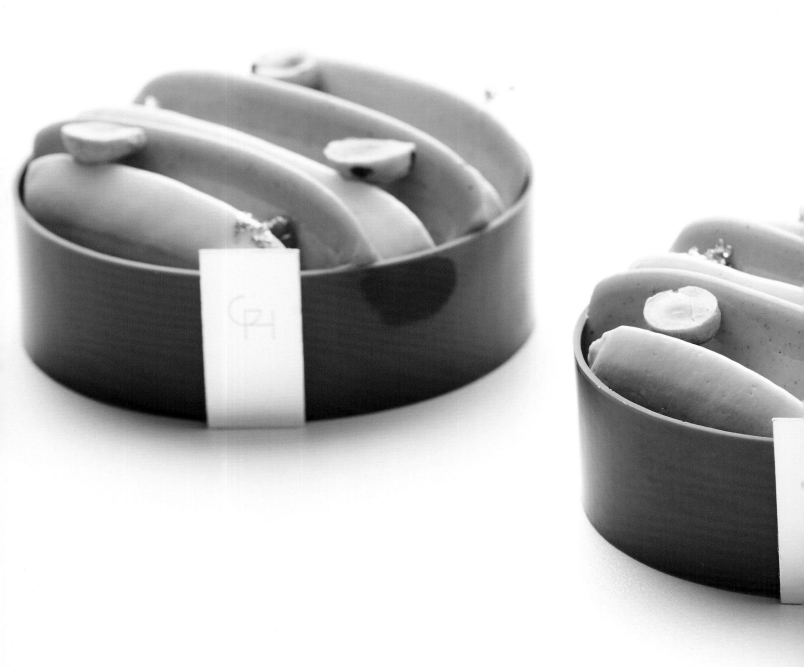

MODELING CHOCOLATE -DARK

모델링 초콜릿-다크

tools & ingredients

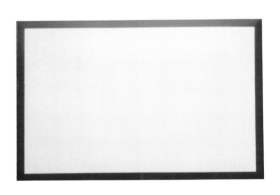  

TOOLS	실팻, 밀대, 원형 커터, 실리콘 패드	Silpat, rolling pin, round cutter, silicone pads
INGREDIENTS	다크초콜릿 180g, 물엿 70g, 설탕 15g, 물 15g, 포도씨유 12g	180g dark chocolate, 70g corn syrup, 15g sugar, 15g water, 12g grapeseed oil

MODELING CHOCOLATE 1

2

3

4

PROCESS 1

2

3-1

3-2

4

Process

모델링 초콜릿	1.	냄비에 물, 설탕, 물엿을 넣고 설탕이 완전이 녹을 때까지 가열한다.
	2.	35℃로 녹인 다크초콜릿에 가열한 시럽과 포도씨유를 순서대로 넣고 혼합한다.
	3.	실팻에 붓고 랩으로 감싼 후 고르게 펼쳐준다.
	4.	상온에서 12시간 세팅시킨 후 사용한다.

공정	1.	모델링 초콜릿을 부드럽게 풀어준 후 밀대를 이용해 얇게 밀어 편다.
	2.	원하는 모양과 사이즈의 커터로 커팅한다.
	3.	실리콘 패드를 이용해 문양을 만든다.
	4.	16~18℃ 온도에서 12시간 정도 굳혀준다.

MODELING CHOCOLATE	1.	In a saucepan, heat water, sugar, and corn syrup until the sugar melt completely.
	2.	Melt dark chocolate to 35℃, and mix with heated syrup and grape seed oil in order.
	3.	Pour over a Silpat and cover with plastic wrap. Spread out evenly.
	4.	Set for 12 hours at an ambient temperature before use.

PROCESS	1.	Knead to soften the modeling chocolate and roll out thin using a rolling pin.
	2.	Cut with cutters of desired shapes and sizes.
	3.	Make imprints using silicone pads.
	4.	Set for about 12 hours at 16~18℃.

MODELING CHOCOLATE -WHITE

모델링 초콜릿-화이트

tools & ingredients

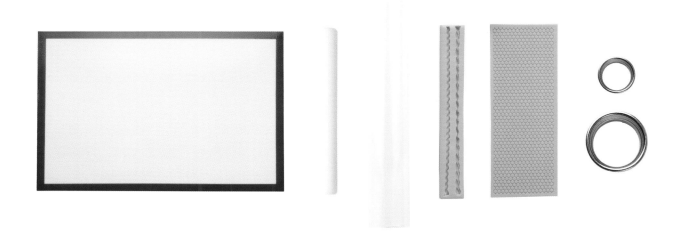

TOOLS	실팻, 밀대, 아크릴파이프, 실리콘 몰드, 실리콘 패드, 원형 커터, 원형 틀	Silpat, rolling pin, acrylic pipe, silicone mold, silicone pad, round cutter, round ring mold
INGREDIENTS	화이트초콜릿 252g, 물엿 70g, 설탕 15g, 물 15g, 포도씨유 10g	252g white chocolate, 70g corn syrup, 15g sugar, 15g water, 10g grapeseed oil

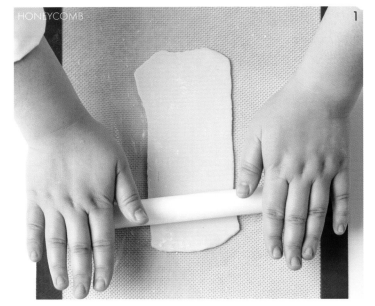

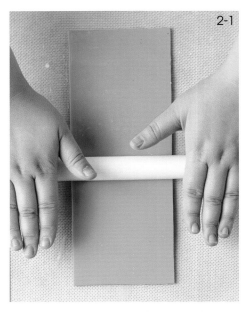

2-1

2-2

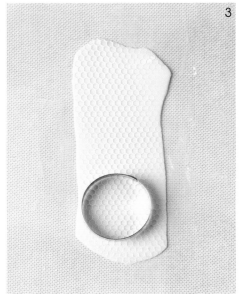

3

4-1

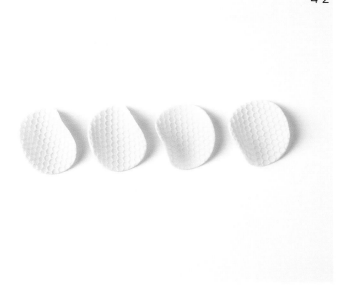

4-2

Process

모델링 초콜릿   1.   화이트초콜릿을 이용해 117p의 방법과 동일하게 모델링 초콜릿을 만든다.

벌집   1.   모델링 초콜릿을 부드럽게 풀어준 후 밀대를 이용해 얇게 밀어 편다.

    2.   실리콘 패드를 이용해 문양을 만든다.

    3.   원하는 모양과 사이즈의 커터로 커팅한다.

    4.   16~18℃ 온도에서 24시간 정도 굳혀준다.

    * 파이프 또는 굴곡이 있는 팬을 이용해 원하는 형태로 성형할 수 있다.

MODELING CHOCOLATE   1.   Make modeling chocolate with white chocolate using the same method on p.117.

HONEYCOMB   1.   Knead to soften the modeling chocolate and roll out thin using a rolling pin.

    2.   Make imprints using a silicone pad.

    3.   Cut with cutters of desired shapes and sizes.

    4.   Set for about 24 hours at 16~18℃.

    * The chocolate can be shaped to desired forms with pipes or curved baking trays.

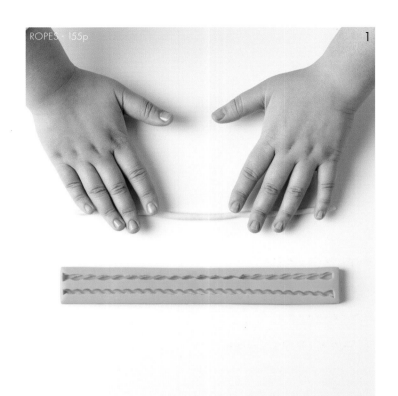

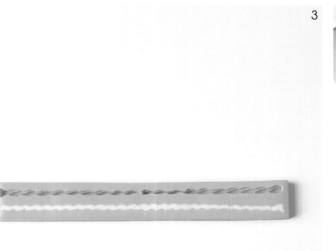

Chocolate

밧줄   1.   모델링 초콜릿을 부드럽게 풀어준 후 막대 형태로 늘려준다.

2.   밧줄 모양의 실리콘 몰드에 채워 넣는다.

3.   냉동고에 넣어 단단히 굳혀준 후 몰드에서 빼낸다.

4.   원하는 모양으로 성형한 후 16~18℃ 온도에서 24시간 정도 굳혀준다.

ROPES   1.   Knead to soften the modeling chocolate and roll out into a long stick.

2.   Fill into the rope-shaped silicone mold.

3.   Freeze to set. Remove from the mold when it's hardened.

4.   Form into the desired shape, and set for about 24 hours at 16~18℃.

PART 02

MOLD & FILM

전문적인 도구를 활용해 나만의 독창적인 디저트를 만들 수 있습니다.

You can make creative desserts of your own using professional tools.

PET MOLD

페트 몰드

tools & ingredients

TOOLS	진공 성형기, 오브제, 열 가소성 시트, 스크래퍼	Vacuum Former, objects, thermoplastic sheets, scraper
INGREDIENTS	커버추어 초콜릿	couverture chocolate

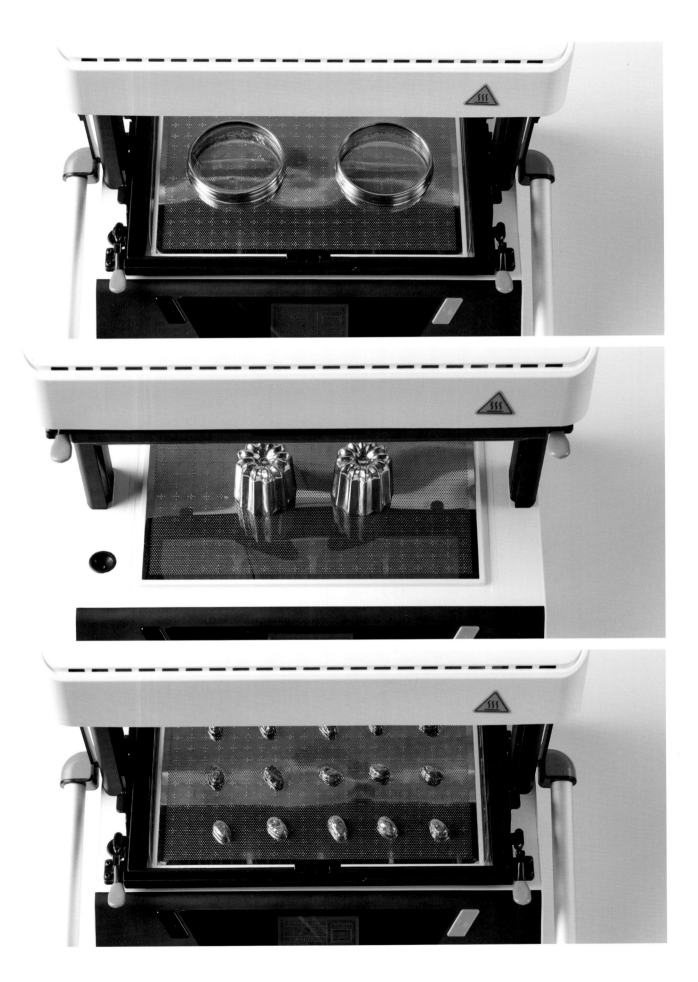

Process

공정 ❶   1.   진공 성형기를 이용해 몰드를 만든다.

                    * 다양한 오브제를 사용해 몰드를 만들 수 있다.

PROCESS ❶   1.   Make molds using the vacuum former.

                    * Various objects can be used to make molds.

공정 ❷   1.   완성된 몰드에 커버추어 초콜릿을 가득 채워준다.

         2.   원하는 두께가 될 때까지 기다려준 후 몰드를 뒤집어 여분의 초콜릿을 털어낸다.

         3.   스크래퍼를 이용해 몰드를 정리한다.

         4.   충분히 수축시킨 후 몰드에서 빼낸다.

PROCESS ❷   1.   Completely fill with couverture chocolate in the finished mold.

            2.   Wait until the desired wall thickness is obtained, then flip over the mold
                  and tap out the excess chocolate.

            3.   Organize the mold using a scraper.

            4.   Remove from the mold after it's crystallized sufficiently.

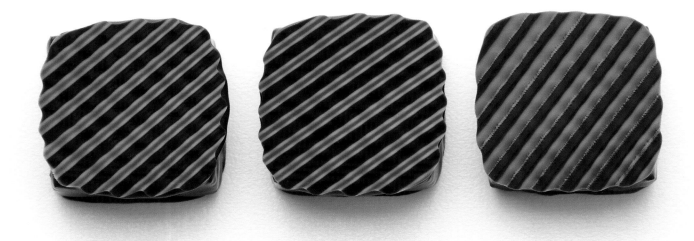

TEXTURED SHEET

텍스처 시트

tools & ingredients

TOOLS	진공 성형기, 열 가소성 시트, 오브제	Vacuum Former, thermoplastic sheets, objects
INGREDIENTS	가나슈 초콜릿, 커버추어 초콜릿	ganache chocolate, couverture chocolate

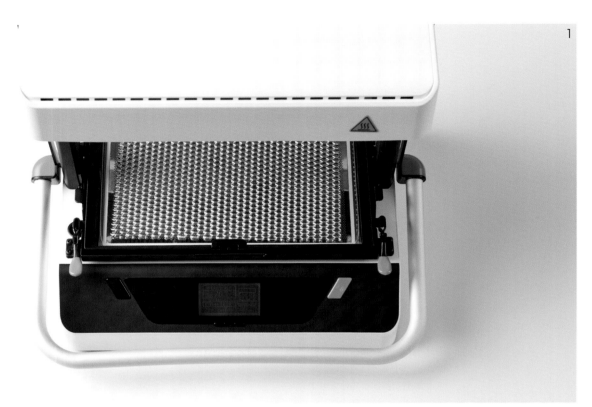

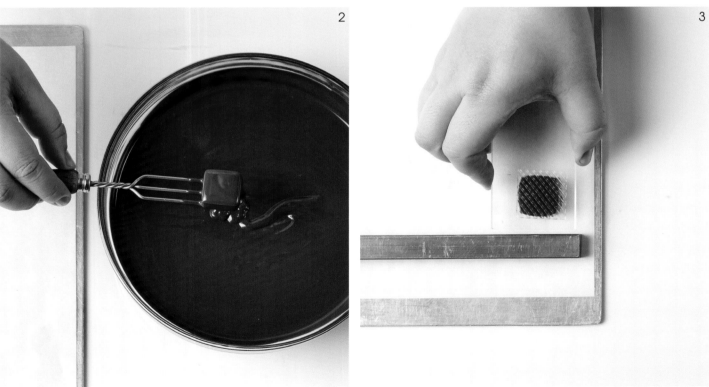

Process

1. 진공 성형기를 이용해 텍스처 시트를 만든 후 원하는 사이즈로 커팅한다.

   * 무늬가 있는 플라스틱, 알루미늄, 실리콘 매트 등 다양한 오브제를 선택할 수 있다.

2. 가나슈를 디핑한다.

3. 커팅한 텍스처 시트를 올려 문양을 만든다.

   * 충분히 수축시킨 후 텍스처 시트를 떼어낸다.

1. Make a textured sheet using the vacuum former and cut the sheet to the preferred size.

   * Various objects can be used, such as plastics with patterns, aluminum, silicone mat, etc.

2. Dip the ganache chocolates.

3. Place the cut textured sheets to make imprints.

   * Crystallize sufficiently, and remove the textured sheets.

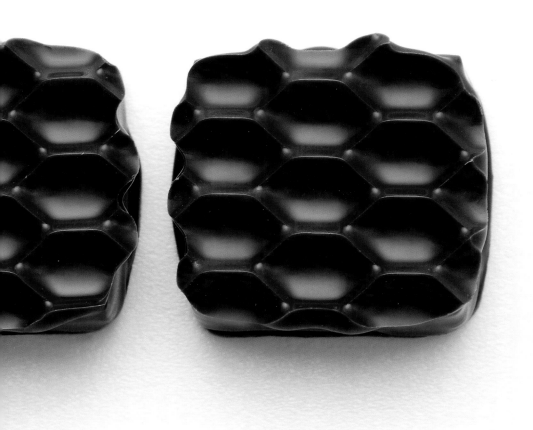

MOLD TECHNIQUE 19
SILICONE MOLD

실리콘 몰드

tools & ingredients

TOOLS	몰드 제작용 실리콘(Sosa), 경화제, 베큠 머신, 우드락, 점토, 밀대, 양면테이프, 오브제, 에어 건, 붓	silicone for mold making(Sosa), hardener, vacuum machine, Woodrock board, clay, rolling pin, double sided tape, objects, air gun, brush
INGREDIENTS	커버추어 초콜릿	couverture chocolate

1

2

3

4-1

4-2

5

6

7-1

7-2

8

9

10

Process

실리콘 몰드

1. 오브제 사이즈에 맞춰 우드락을 재단한다.

2. 점토를 얇게 밀어 편 후 우드락과 동일한 모양으로 재단한다.

3. 재단한 점토를 우드락 위에 올려준다.

4. 오브제에 맞춰 실리콘을 부을 높이를 결정한 후 필름을 재단한다. 양면테이프를 이용해 우드락에 고정시킨다.

5. 점토 위에 오브제를 고정시킨다.

6. 실리콘과 경화제를 비율에 맞춰 계량한 후 충분히 혼합한다.

7. 베큠 머신을 이용해 탈포한다.

8. 준비한 오브제 위에 실리콘을 부어준다.

9. 부을 때 생기는 기포는 에어 건을 이용해 정리한다.

10. 상온에서 충분히 경화시킨 후 실리콘 몰드를 오브제와 분리한다.

    * 온도에 따라 경화 시간에 차이가 생길 수 있다. 겨울철 실내 온도가 낮을 경우 경화 시간이 길어지며, 반대로 온도가 높을 경우에는 경화 시간이 짧아질 수 있다.

SILICON MOLD

1. Cut the woodrock board according to the size of the objects.

2. Roll out the clay thin, and cut it into the same shape as the woodrock board.

3. Place the cut clay over the woodrock board.

4. Determine the height of the silicone to pour according to the object and cut a film. Use double sided tape to fix it to the woodrock board.

5. Fix the object on top of the clay.

6. Weigh the silicone and hardener according to the ratio and mix thoroughly.

7. Defoam using the vacuum machine.

8. Pour the silicone over the prepared objects.

9. Remove the air bubbles generated during pouring with an air gun.

10. Harden sufficiently in ambient temperature, and remove the silicone mold from the object.

    * Hardening time may differ by temperature. It will take longer when the room temperature is low during winter, and conversely, when the temperature is high, hardening time may be shortened.

FEATHER

깃털	1.	완성된 실리콘 몰드에 템퍼링한 커버추어 초콜릿을 채워준다. 이때 깃털 사이사이 주름 부분까지 초콜릿이 잘 채워지도록 붓을 이용해 꼼꼼히 작업한다.
	2.	투명 필름을 얹고 뒤집어준 후 손으로 눌러 기포를 제거하면서 몰드 안에 초콜릿이 잘 밀착되도록 한다.
	3.	초콜릿이 완전히 굳기 전에 실리콘 몰드를 조심스럽게 떼어낸다.
	4.	굴곡이 있는 팬에 얹어 모양을 만든다.
	5.	뾰족한 도구를 이용해 깃털 라인을 정리한다.
	6.	충분히 굳힌 후 필름을 떼어낸다.

FEATHER	1.	Fill the finished silicone mold with tempered couverture chocolate. Make sure to fill the chocolate in all the ridges of the feather using a brush.
	2.	Cover with a clear plastic film and turn it upside down. Press with hand to remove any air bubbles, and make sure the chocolate adheres well inside the mold.
	3.	Remove the silicone mold carefully before the chocolate completely crystallizes.
	4.	Place on a curved baking tray to form its shape.
	5.	Organize the lines of the feather with a pointy tool.
	6.	Remove the film after the chocolate is crystallized enough.

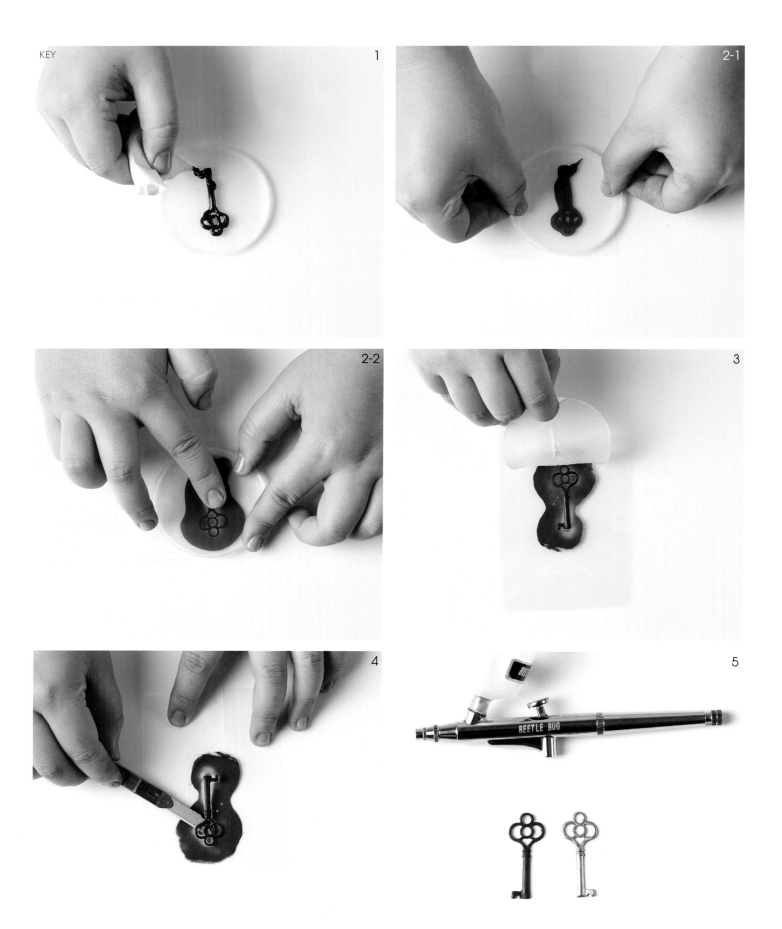

KEY

열쇠

1. 완성된 실리콘 몰드에 템퍼링한 커버추어 초콜릿을 파이핑한다.

2. 투명 필름을 얹고 뒤집어준 후 손으로 눌러 기포를 제거하면서 몰드 안에 초콜릿이 잘
    밀착되도록 한다.

3. 초콜릿이 완전히 굳기 전에 실리콘 몰드를 조심스럽게 떼어낸다.

4. 충분히 굳힌 후 나이프를 이용해 초콜릿 장식물을 필름에서 떼어낸다.

5. 원하는 경우 스프레이 건을 이용해 조색할 수 있다.

KEY

1. Pipe tempered couverture chocolate in the finished silicone mold.

2. Cover with a clear plastic film and turn it upside down. Press with hand to remove
    any air bubbles, and make sure the chocolate adheres well inside the mold.

3. Remove the silicone mold carefully before the chocolate completely crystallizes.

4. Crystallize sufficiently and remove the chocolate decoration from the film
    with a knife.

5. If desired, color may be applied using a spray gun.

GELATIN MOLD

젤라틴 몰드

tools & ingredients

TOOLS	베큠 머신, 우드락, 점토, 밀대, 양면테이프, 오브제, 붓	vacuum machine, Woodrock board, clay, rolling pin, double sided tape, objects, brush
INGREDIENTS	젤라틴 20g, 물 60g	20g gelatin, 60g water

1
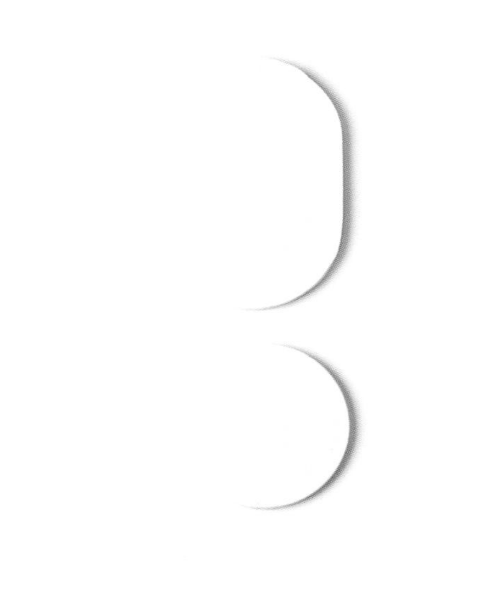

2
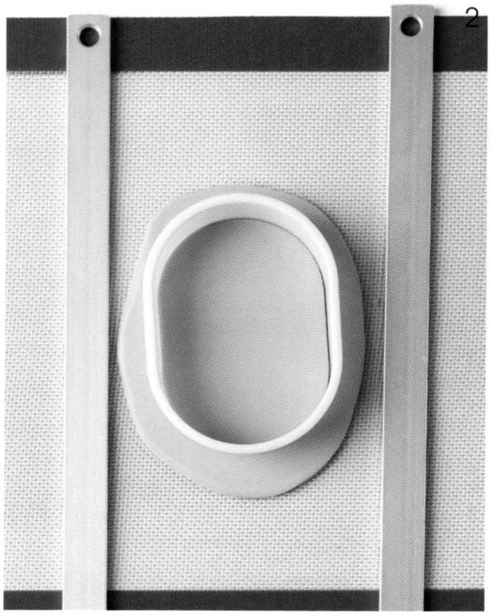

3
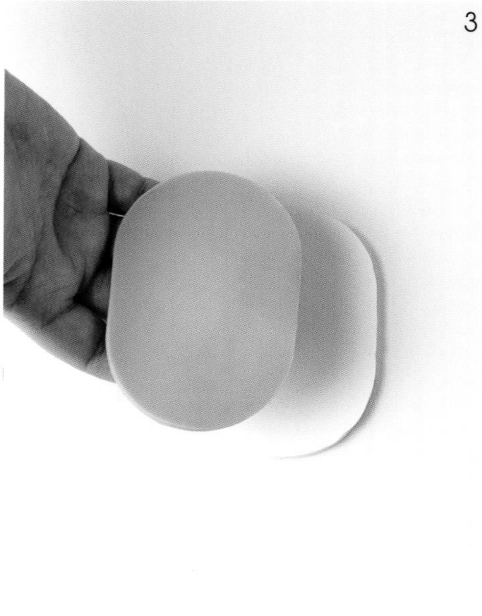

4-1
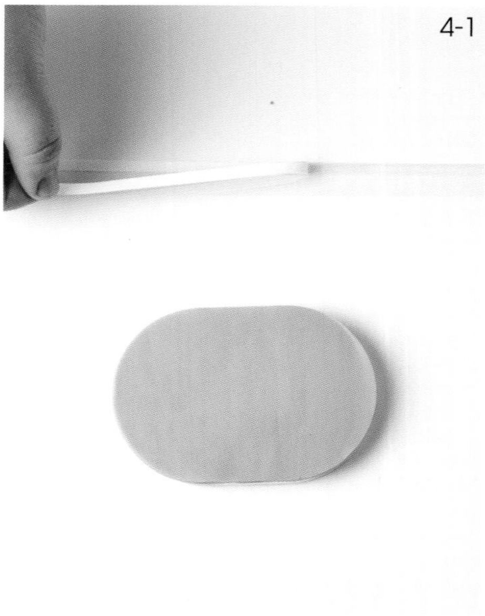

4-2
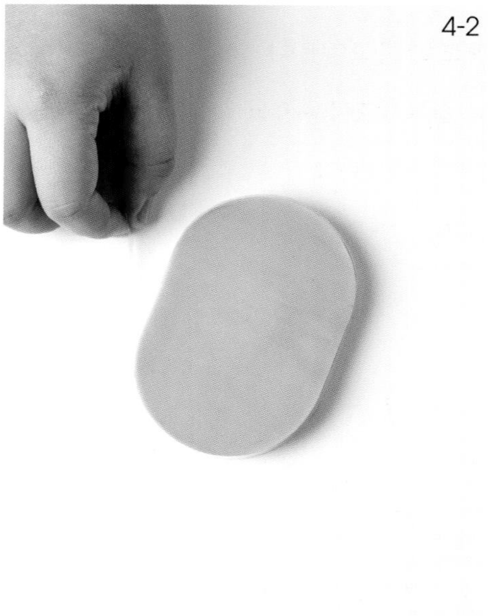

5
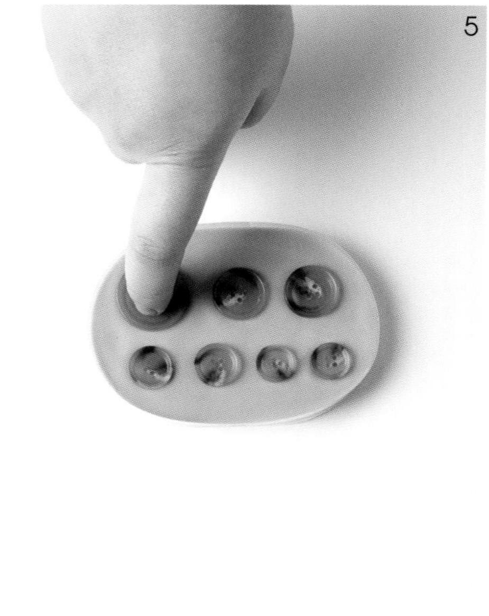

6
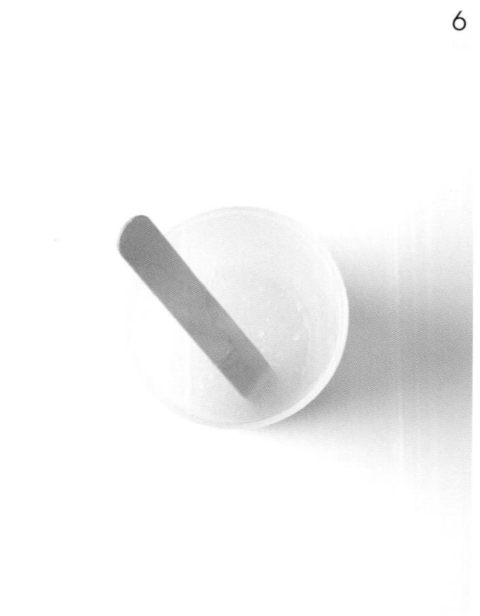

7-1
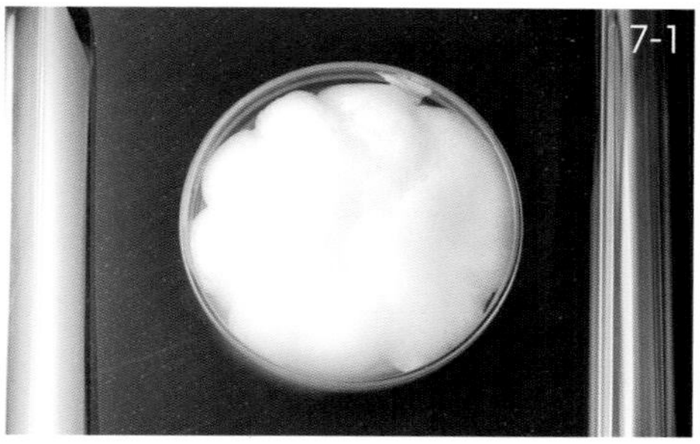

7-2
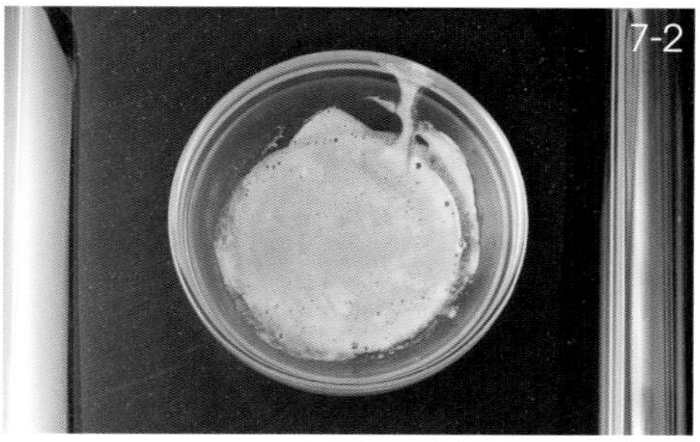

8
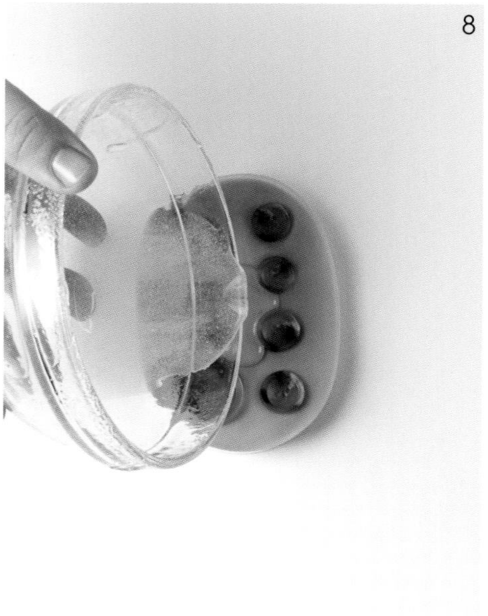

Process

젤라틴 몰드

1. 오브제 사이즈에 맞춰 우드락을 재단한다.

2. 점토를 얇게 밀어 편다.

3. 우드락과 동일한 모양으로 재단한 후 우드락 위에 올려준다.

4. 오브제에 맞춰 실리콘을 부을 높이를 결정한 후 필름을 재단한다. 양면테이프를 이용해 우드락에 고정시킨다.

5. 점토위에 오브제를 고정시킨다.

6. 젤라틴과 물을 혼합한다. 이때 물의 온도는 50℃로 맞춰준다.

7. 베큠 머신을 이용해 탈포한다.

8. 준비한 오브제 위에 부어준다.

GELATIN MOLD

1. Cut the woodrock board according to the size of the objects.

2. Roll out the clay thin.

3. Cut the clay into the same shape as the woodrock board, and place it on the board.

4. Determine the height of the silicone to pour according to the objects and cut a film. Use double sided tape to fix it to the woodrock board.

5. Fix the objects on top of the clay.

6. Combine gelatin and water. Make sure to set the water temperature to 50℃.

7. Defoam using the vacuum machine.

8. Pour over the prepared objects.

9

10

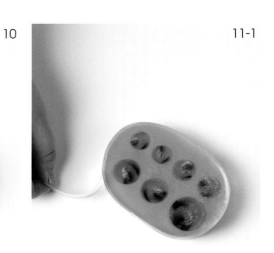

11-1

11-2

BUTTON

1

2-1

2-2

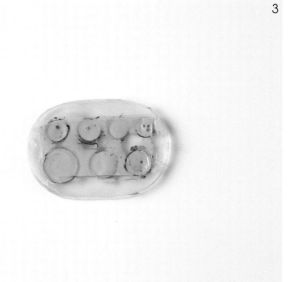

3

9. 부을 때 생기는 기포는 토치를 이용해 정리한다.

10. 냉장고에 넣어 단단하게 굳힌다.

11. 오브제와 분리한다.

단추

1. 완성한 젤라틴 몰드에 템퍼링한 다크 커버추어 초콜릿을 붓으로 터치한다.

2. 초콜릿이 완전히 굳으면 블론드 커버추어 초콜릿을 가득 채워준 후 표면을 매끄럽게 정리한다.

3. 충분히 수축시킨 후 몰드에서 빼낸다.

9. Remove the air bubbles generated during pouring with a blow torch.

10. Store in the fridge to harden.

11. Separate from the objects.

BUTTONS

1. Brush lightly with tempered dark couverture chocolate on the finished gelatin mold.

2. When the chocolate is completely crystallized, fill it with Blond couverture chocolate and smooth out the surface.

3. Crystallize sufficiently before removing from the mold.

DESIGN FILM

디자인 필름

tools

커팅기, 커팅용 필름

cutting machine, films for cutting

Process

디자인 필름   1.   커팅기를 이용해 디자인한 도안을 커팅한다.

            * 사용 목적에 맞게 필름의 재질, 두께를 결정할 수 있다.

DESIGN FILM   1.   Cut the designed pattern using a cutting machine.

            * Material and thickness of the film can be determined according to the purpose of use.

1

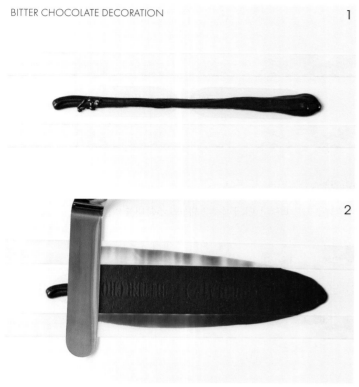

2

3

4

5

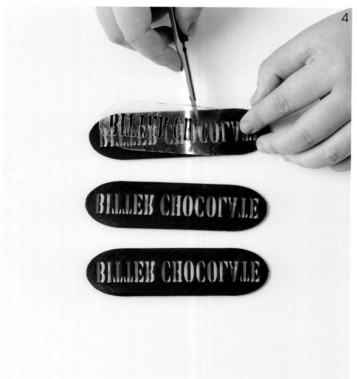

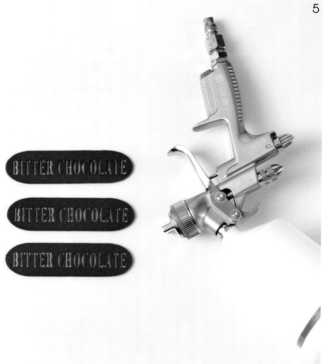

비터초콜릿
장식물

1. 커팅한 필름 위에 템퍼링한 커버추어 초콜릿을 적당량 부어준다.

2. 스패츌러를 이용해 초콜릿을 1mm 두께로 얇게 펴준다. 이때 실리콘 재질의 두께 바를 이용하면 일정한 두께로 밀어 펼 수 있다.

3. 초콜릿이 굳기 시작하면 조심스럽게 필름을 들어올린다.

4. 충분히 수축시킨 후 필름을 떼어내고 냉동고에 30분 동안 넣어준다.

5. 커버추어 초콜릿과 카카오 버터를 1:1 비율로 혼합한 후 차가워진 초콜릿 표면해 분사해 벨벳 같은 느낌을 표현한다.

BITTER
CHOCOLATE
DECORATION

1. Pour a moderate amount of tempered couverture chocolate over the cut film.

2. Spread out thin to 1mm with a spatula. Using silicone confectionery bars will help to spread out evenly.

3. When the chocolate starts to crystallize, carefully lift the film.

4. Crystallize sufficiently and remove the film. Store in the freezer for 30 minutes.

5. Combine couverture chocolate and cacao butter to 1:1 ratio. Spray on the surface of the chilled chocolate decoration to express velvet-like texture.

PART 03

SUGAR PASTE & MERINGUE

머랭 장식물은 당도와 식감을 더해주는 요소로 사용할 수 있습니다. 레몬 타르트 토핑으로도 주로 사용하는 이탈리안 머랭은 당도를 더해주면서 부드럽고 폭신폭신한 식감으로 레몬 크림의 산미를 중화시켜주며, 건조시킨 머랭은 디저트에 크런치한 식감을 더해줍니다. 또한 머랭에 과일 제스트나 견과류 분말을 활용해 고유의 풍미를 입혀줄 수도 있습니다.

Meringue decorations can be used as an element to add sweetness and texture. The Italian meringue, usually used as a topping for a lemon tart, neutralizes the acidity of lemon cream with a soft and fluffy texture while it adds sweetness. Dried meringue adds a crunchy texture to the desserts. Also, a distinct flavor can be added by using the zest of fruits or nut powders.

SUGAR PASTE

슈거 페이스트

tools & ingredients

TOOLS	실팻, 모양 커터, 밀대	Silpat, shape cutters, rolling pin
INGREDIENTS	**슈거 페이스트** 슈거파우더 250g, 레몬즙 5g, 젤라틴매스(젤라틴 4g, 물 20g) 24g	**SUGAR PASTE** 250g powdered sugar, 5g lemon juice, 24g gelatin mass (4g gelatin, 20g water)
	로열아이싱 슈거파우더 50g, 달걀흰자 10g	**ROYAL ICING** 50g powdered sugar, 10g egg whites

1

2

3

4

1

2

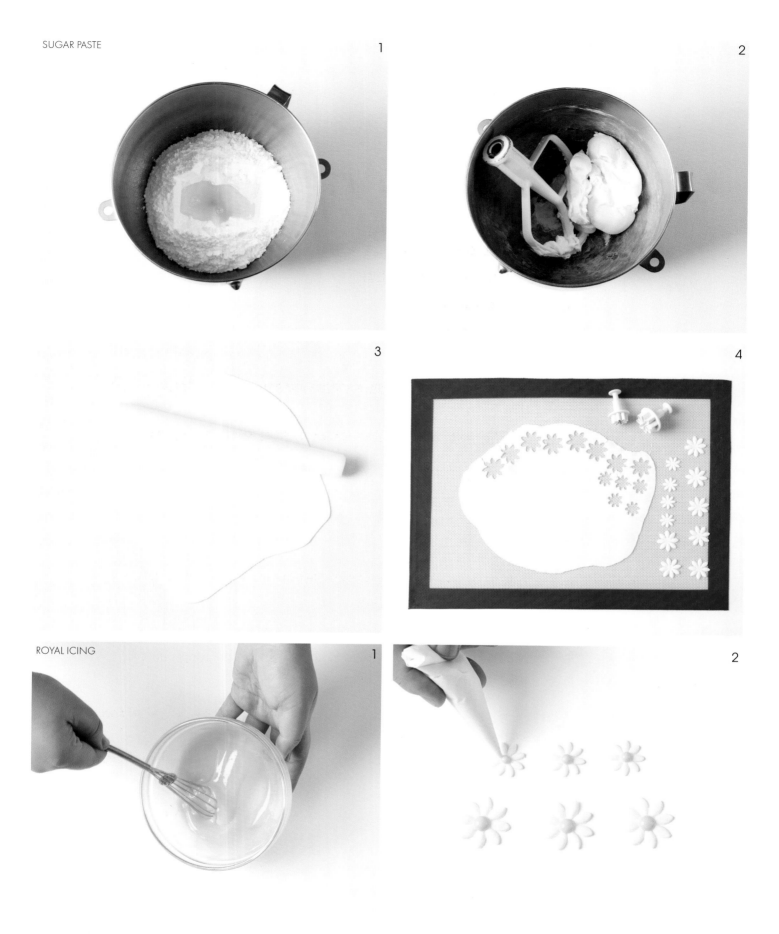

Process

슈거 페이스트
1. 슈거파우더, 레몬즙, 녹인 젤라틴매스를 믹싱볼에 담아준다.
2. 반죽이 한 덩어리가 될 때까지 믹싱한다.
3. 반죽을 최대한 얇게 밀어 편다.
4. 원하는 모양의 커터로 커팅한 후 상온에서 하루 동안 건조시켜 사용한다.

로열 아이싱
1. 슈거파우더에 달걀흰자를 넣고 혼합한 후 휘핑한다. 이때 천연 식용 색소를 사용해 원하는 색으로 조색할 수 있다.
2. 건조시킨 슈거페이스트 반죽 위에 원하는 모양으로 파이핑한다. 상온에서 하루 동안 충분히 건조시킨다.

SUGAR PASTE
1. In a mixing bowl, add powdered sugar, lemon juice, and melted gelatin mass.
2. Mix until the dough forms a ball.
3. Roll out the dough as thin as possible.
4. Cut using desired cutters, and dry for a day at an ambient temperature.

ROYAL ICING
1. Stir in egg whites into powdered sugar and whip. At this time, natural food colors can be added to make desired colors.
2. Pipe on the dried sugar paste dough in desired shapes. Dry sufficiently in ambient temperature for a day.

MERINGUE TECHNIQUE 23
MERINGUE

머랭

tools & ingredients

TOOLS	테프론 시트, 각봉, 자, 비스퀴 스패출러	Teflon sheets, aluminum confectionery bars, ruler, biscuit spatula
INGREDIENTS	달걀흰자 50g, 설탕 50g, 슈거파우더 50g	50g egg whites, 50g sugar, 50g powdered sugar

MERINGUE 1

2

3

FORMATION ❶ 4

FORMATION ❷ 5

Martellato
www.martellato.com
SPATOLA BISCUIT - KNIFE BISCUIT
SPB

6

Process

머랭	1.	냄비에 달걀흰자와 설탕을 넣고 65℃까지 가열한다.
	2.	머랭이 하얗게 올라오고 힘 있는 상태가 될 때까지 휘핑한다.
	3.	체 친 슈거파우더를 넣고 혼합한다.
성형 ❶	4.	원형 깍지를 이용해 원뿔 모양으로 파이핑한다.
성형 ❷	5.	테프론시트 위에 얇게 밀어 편다. 이때 각봉을 이용하면 균일한 두께로 밀어 펼 수 있다.
	6.	70℃ 오븐에서 약 1시간 동안 건조시킨다.

TIP. 건조시킨 머랭을 수분이 많은 케이크 위에 올릴 경우 쉽게 눅눅해질 수 있으므로 스프레이 건을 이용해 카카오버터를 표면에 도포한 후 사용할 수 있다.

MERINGUE	1.	In a saucepan, add egg whites and sugar. Heat until the temperature reaches 65℃.
	2.	Whip until the meringue turns white and obtains volume, with a firm structure.
	3.	Combine sifted powdered sugar.
FORMATION ❶	4.	Pipe in cone shapes using a round tip.
FORMATION ❷	5.	Spread out thin on a Teflon sheet. Using bars will help to roll out to even thickness.
	6.	Dry for about 1 hour in an oven at 70℃.

TIP. When dried meringue is used on a cake with lots of moisture, it can easily get soggy. Therefore, it can be used after its been sprayed with cacao butter on the surface.

NUTS & BEANS

메뉴에 향과 식감을 더해줄 수 있습니다. 로스팅한 견과류 또는 빈을 그대로 사용하거나, 캐러멜라이즈할 수 있으며 튀일이나 누가틴으로도 활용할 수 있습니다.

These can add aroma and texture to the menu. Roasted nuts and beans can be used as-is or caramelized and can be used as tuile or nougatine.

NUTS & BEANS

너트 & 빈

CRUNCHY CACAO

크런치 카카오

ingredients

———

물 42g

설탕 125g

카카오닙 150g

42g water

125g sugar

150g cacao nibs

Process

1. 카카오닙은 160℃ 오븐에서 3분간 데워준다.

2. 냄비에 물과 설탕을 넣고 118~121℃까지 가열한다.

3. 따뜻한 상태의 카카오닙을 넣고 섞어준다.

4. 시럽이 하얗게 재결정화 상태가 될 때까지 골고루 저어준다.

5. 실팻에 넓게 펼친 후 충분히 식힌다.

1. Warm the cacao nibs in an oven at 160℃ for 3 minutes.

2. In a saucepan, heat water and sugar to 118~121℃.

3. Stir in the warm cacao nibs.

4. Mix well until the syrup recrystallizes and turns white.

5. Spread wide over a Silpat and cool sufficiently.

CARAMELIZED HAZELNUTS

캐러멜라이즈 헤이즐넛

ingredients

물 19g

설탕 64g

헤이즐넛 200g

카카오버터 8g

19g water

64g sugar

200g hazelnuts

8g cacao butter

Process

1. 헤이즐넛은 140℃ 오븐에서 약 8분간 구워준다.

2. 냄비에 물과 설탕을 넣고 118~121℃까지 가열한다.

3. 따듯한 상태의 헤이즐넛을 넣고 섞어준다.

4. 시럽이 하얗게 재결정화 상태가 될 때까지 계속해서 저어준다.

5. 다시 불에 올려 골고루 저어주며 계속해서 가열한다.

6. 진한 캐러멜 색이 나면 불에서 내린 후 카카오버터를 넣고 고르게 섞어준다.

7. 실팻에 부어 펼친 후 완전히 식기 전에 한 알씩 떼어낸다.

1. Roast the hazelnuts in an oven at 140℃ for about 8 minutes.

2. In a saucepan, heat water and sugar to 118~121℃.

3. Stir in the warm hazelnuts.

4. Continue to stir until the syrup recrystallizes and turns white.

5. Put it back over the heat, and continue to stir.

6. When the color becomes a dark caramel color, remove from the heat, and mix well with cacao butter.

7. Pour over a Silpat and separate the nuts one by one before it's completely cooled.

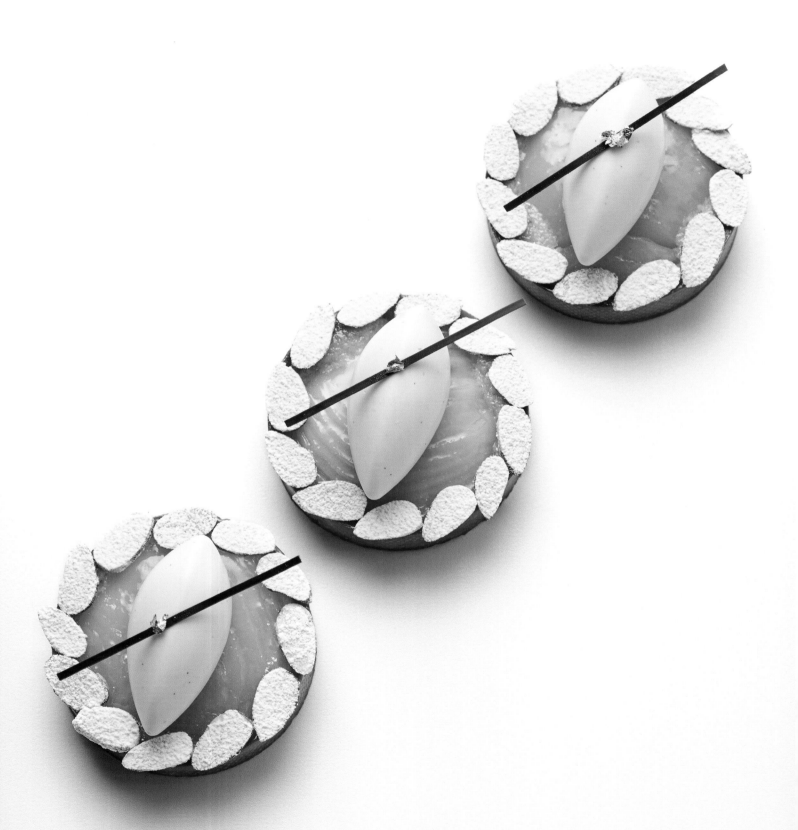

TECHNIQUE 27

SNOW ALMOND

스노우 아몬드

ingredients

슬라이스 아몬드
달걀흰자
데코스노우

sliced almonds
egg whites
decosnow

Process

1.   슬라이스 아몬드 표면에 달걀흰자를 발라준다.

1.   Brush egg whites on the surface of the sliced almonds.

2. 데코스노우를 도포한 후 50℃ 오븐에서 약 2시간 동안 건조시킨다.

2. Cover with decosnow and dry for about 2 hours in an oven at 50℃.

GOLDEN NUTS

골든 너트

tools & ingredients

TOOLS	붓, 핀셋	brush, tweezers
INGREDIENTS	로스팅한 견과류, 식용금박	roasted nuts, edible gold leaves

3-1

3-2

Process

1. 식용금박 위에 로스팅한 견과류를 올린다.

2. 붓을 이용해 견과류를 굴려준다.
   * 핀셋을 이용해 여분의 식용금박이 딸려오지 않도록 한다.

3. 식용금박을 견과류 표면에 균일하게 입혀준다.

1. Place a roasted nut on a sheet of edible gold leaf.

2. Roll the nut using a brush.
   * Make sure to use tweezers to avoid excess gold leaf from adhering.

3. Wrap the gold leaf evenly on the surface of the nut.

PART 05

TUILE &
NOUGATINE &
CRISPY

메뉴에 바삭한 식감과 고소한 맛을 더해줄 수 있습니다. 메뉴에 어울리는 견과류 또는 빈 등을 활용해
다양하게 변형할 수 있습니다.

A crispy texture and nutty flavor can be added to the menu. It can be modified
in various ways by using nuts and beans that go well with the menu.

ALMOND NOUGATINE

아몬드 누가틴

tools & ingredients

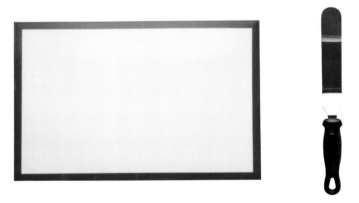

TOOLS	실팻, 스패출러	Silpat, spatula
INGREDIENTS	슬라이스 아몬드 120g, NH펙틴 2.5g, 슈거파우더 100g, 버터 83g, 물엿 33g	120g sliced almonds, 2.5g pectin NH, 100g powdered sugar, 83g butter, 33g corn syrup

Process

1. 믹싱볼에 슬라이스 아몬드, 슈거파우더, NH펙틴을 넣고 비터를 이용해 골고루 섞는다.

2. 뜨겁게 데운 버터와 물엿을 순서대로 넣고 혼합한다.

3. 트레이에 붓고 상온에서 12시간 휴지시킨다.

4. 두 장의 실팻 사이에 휴지시킨 누가틴 반죽을 넣고 2mm 두께로 밀어 편다.

5. 160℃ 오븐에서 약 12분간 굽는다.

1. In a mixing bowl, add sliced almonds, powdered sugar, and pectin NH. Mix well with a paddle attachment.

2. Stir in hot butter and corn syrup in order.

3. Pour in a tray and let rest for 12 hours at an ambient temperature.

4. Place the nougatine dough between two sheets of Silpat and roll out to 2mm thickness.

5. Bake in an oven at 160°C for about 12 minutes.

TUILE TECHNIQUE 30

SESAME TUILE

깨 튀일

tools & ingredients

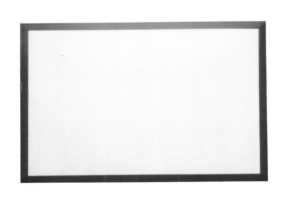

TOOLS	실팻, 스패출러, 붓	Silpat, spatula, brush
INGREDIENTS	참깨 70g, 검정깨 70g, 볶은 현미 20g, 박력분 30g, 달걀흰자 80g, 설탕 56g, 버터 40g, 화이트 코팅초 콜릿, 카카오버터	70g sesame seeds, 70g black sesame seeds, 20g roasted brown rice, 30g cake flour, 80g egg whites, 56g sugar, 40g butter, white compound chocolate, cacao butter

Process

1. 달걀흰자, 설탕, 체 친 박력분을 섞어준다.

2. 1에 참깨, 검정깨, 볶은 현미를 넣고 섞어준다.

3. 55~60℃로 녹인 버터를 혼합한다.

4. 스패출러를 이용해 실팻에 고르게 펼친 후 180℃로 예열된 오븐에서 10분간 굽는다.

5. 오븐에서 나온 직후 튀일 표면에 녹인 카카오버터를 얇게 바른다.

6. 완전히 식힌 후 튀일 뒷면에 녹인 화이트 코팅초콜릿을 얇게 바른다.

   * 5번과 6번 공정은 튀일이 쉽게 눅눅해지는 것을 방지해준다.

1. Mix egg whites, sugar, and sifted cake flour.

2. Stir in sesame seeds, black sesame seeds, and roasted brown rice to 1.

3. Mix with butter melted to 55-60℃.

4. Spread evenly on a Silpat with a spatula, and bake for 10 minutes in an oven preheated to 180℃.

5. As soon as the tuile is out of the oven, brush a thin layer of cacao butter on the surface.

6. Cool completely, and brush a thin layer of white compound chocolate on the backside of the tuile.

   * Procedures 5 and 6 prevent the tuile from getting soggy easily.

CRISP TECHNIQUE 31

CRISPY CACAO

크리스피 카카오

tools & ingredients

TOOLS	테프론시트, 밀대	Teflon sheets, rolling pin
INGREDIENTS	카카오닙 125g, 설탕 100g	125g cacao nibs, 100g sugar

Tuile & Nougatine & Crispy

Process

1. 카카오닙은 160℃ 오븐에서 3분간 데워준다. 냄비에 설탕을 넣고 가열하여 캐러멜을 만든다.

2. 따뜻한 상태의 카카오닙을 넣고 고르게 섞어준다.

3. 테프론시트 위에 부어준다.

4. 테프론시트 한 장을 더 올린 후 식기 전에 재빨리 얇게 펼쳐준다.

1. Warm the cacao nibs in an oven at 160℃ for 3 minutes. In a saucepan, heat sugar and make a caramel.

2. Add the warm cacao nibs and stir to mix evenly.

3. Pour over a Teflon sheet.

4. Cover with another Teflon sheet, and roll out quickly before it cools down.

PART 06

ETC

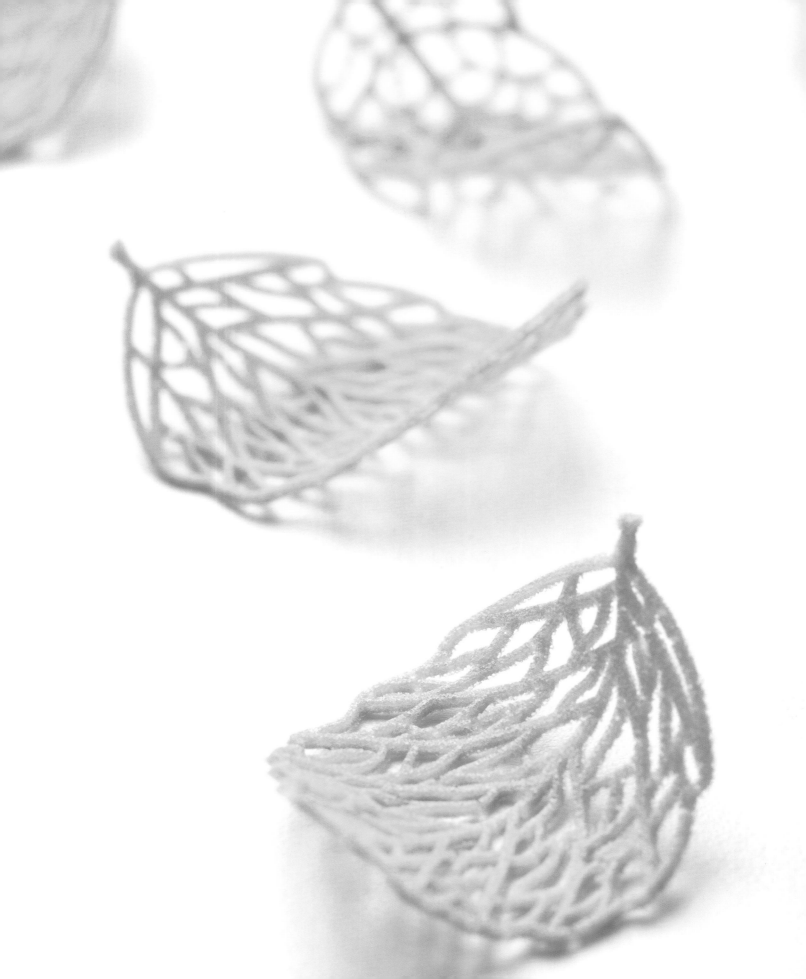

CIGARETTE DOUGH

시가렛 도우

tools & ingredients

TOOLS	실리콘 몰드, 스패출러	silicone mold, spatula
INGREDIENTS	달걀흰자 50g, 슈거파우더 60g, 박력분 36g, 버터 30g	50g egg whites, 60g powdered sugar, 36g cake flour, 30g butter

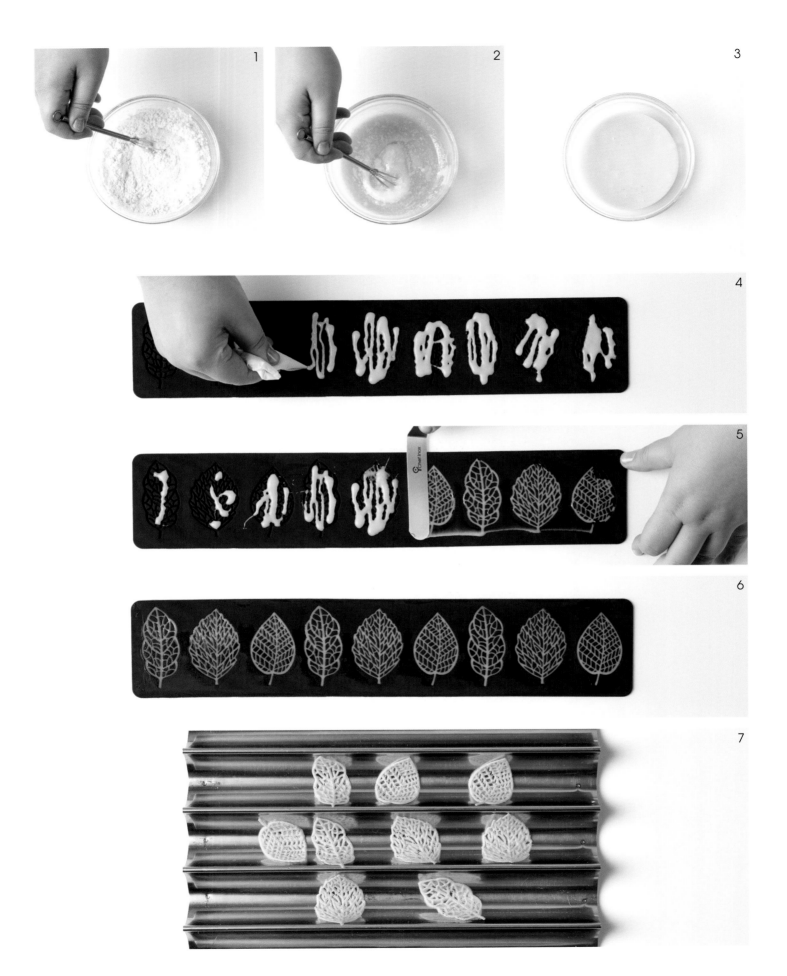

Process

1. 달걀흰자에 슈거파우더와 체 친 박력분을 순서대로 혼합한다.

2. 55℃로 녹인 버터를 넣고 섞어준다.

3. 냉장고에서 12시간 휴지시킨다.

4. 나뭇잎 모양 실리콘 몰드에 시가렛 반죽을 파이핑한다.

5. 스패츌러를 이용해 반죽을 밀어 펴 채워준다.

6. 170℃ 오븐에 약 10분간 구워준다.

7. 뜨거운 상태에서 재빨리 몰드에서 떼어낸 후 식기 전에 굴곡이 있는 팬에 올려 모양을 만든다.

   * 수분과 닿는 순간 쉽게 눅눅해지기 때문에 즉시 서빙하는 디저트에 적합하다.

1. In egg whites, mix powdered sugar and sifted cake flour in order.

2. Stir in butter melted to 55℃.

3. Rest in the fridge for 12 hours.

4. Pipe the cigarette dough on the leaf-shaped silicone mold.

5. Use a spatula to spread the dough thinly to fill in the ridges.

6. Bake for about 10 minutes in an oven at 170℃.

7. Remove from the mold quickly before it cools down, and put on a curved baking tray to form the shape.

   * The cookie gets soggy easily as soon as it comes to contact with moisture. Therefore, it's suitable for desserts serving immediately.

CRYSTAL ISOMALT

크리스탈 이소말트

tools & ingredients

———

TOOLS	실팻	Silpat
INGREDIENTS	이소말트	Isomalt

Process

1. 두 장의 실팻 사이에 이소말트를 얇게 펼쳐준다.

2. 160℃ 오븐에서 약 15분간 열을 가한다.

3. 완전히 식힌 후 실팻에서 떼어낸다.

   * 습기에 약하기 때문에 즉시 서빙하는 디저트에 적합하다.

1. Spread a thin layer of isomalt between two sheets of Silpat.

2. Heat for about 15 minutes in an oven at 160℃.

3. Cool completely and remove from Silpat.

   * Because it is vulnerable to humidity, it's suitable for desserts serving immediately.

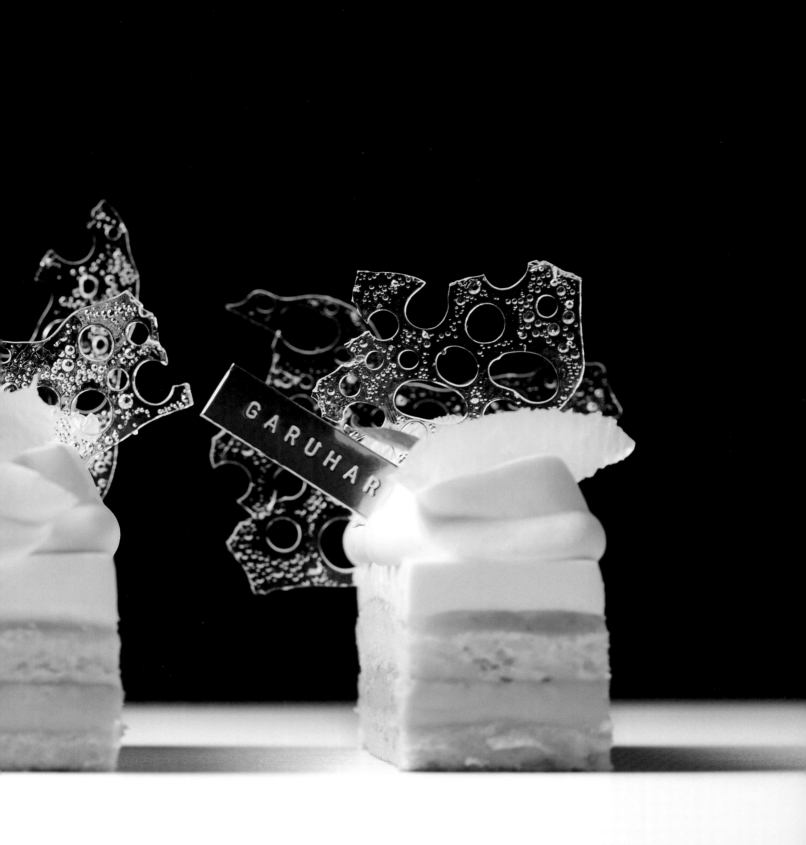

MARZIPAN

마지팬

tools & ingredients

———

TOOLS	밀대, 마지팬 성형 도구	rolling pin, marzipan shaping tools
INGREDIENTS	마지팬, 슈거파우더, 천연 식용 색소	marzipan, powdered sugar, natural food colors

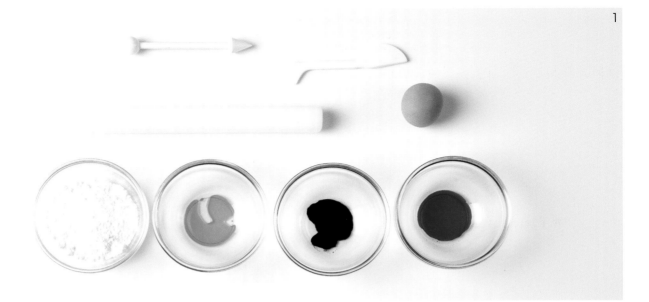

Process

1. 마지팬, 덧가루용 슈거파우더, 천연 식용 색소를 준비한다.

2. 천연 식용 색소를 이용해 마지팬을 원하는 색으로 조색한다. 이때 손에 달라붓지 않도록
   슈가파우더를 덧가루 용도로 사용한다.

3. 원하는 모양으로 성형한다.

1. Prepare marzipan, powdered sugar for dusting, and natural fool colors.

2. Color the marzipan with natural food colors to make the desired color. During
   the process, use powdered sugar for dusting to prevent sticking to the hands.

3. Form into preferred shapes.

BUTTER DECORATION

버터 장식물

tools & ingredients

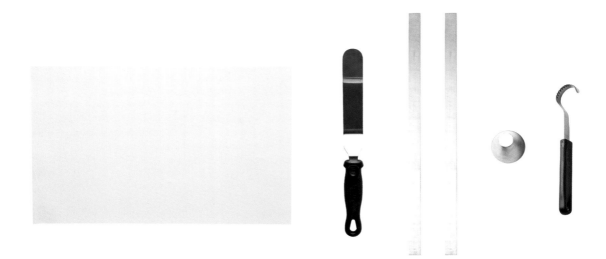

TOOLS	텍스처 시트, 스패츌러, 각봉, 원형 커터, 데커레이션 나이프	textured sheet, spatula, aluminum confectionery bars, round cutter, decoration knife
INGREDIENTS	버터	butter

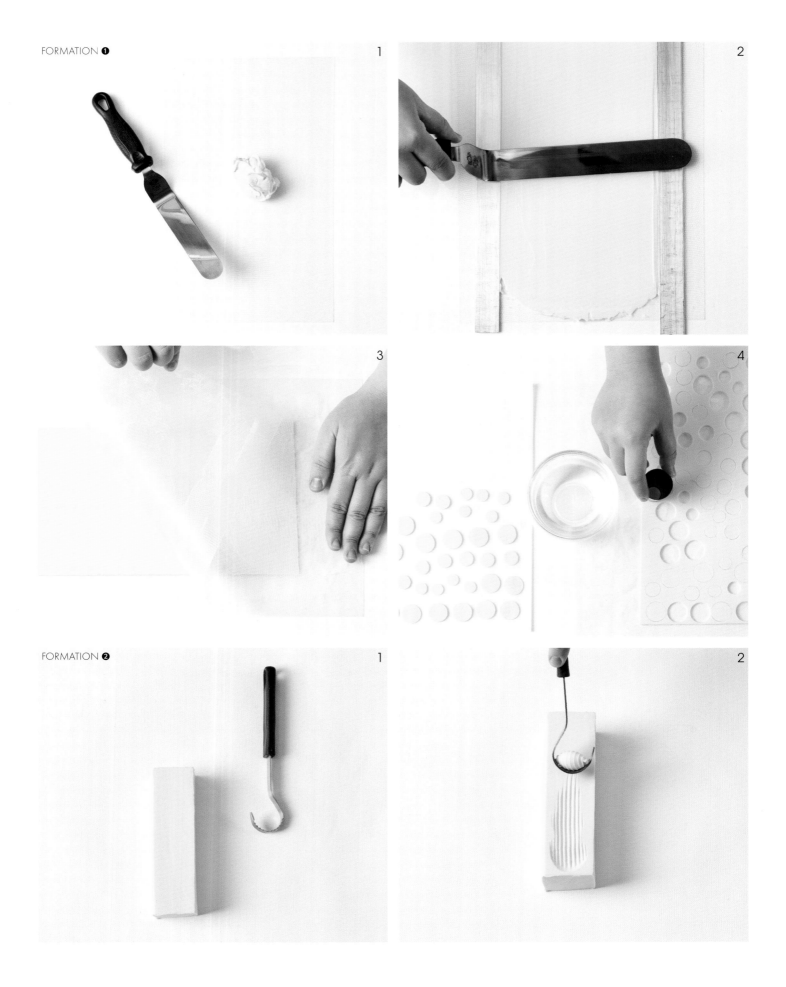

FORMATION ❶

1

2

3

4

FORMATION ❷

1

2

Process

성형 ❶

1. 버터는 상온 상태로 준비한다.

2. 텍스처 시트에 버터를 2mm 두께로 펼친 후 냉장고에서 굳혀준다.

3. 버터가 단단하게 굳으면 시트를 떼어낸다.

4. 원하는 모양으로 커팅한다.
    * 커터를 사용할 경우 따뜻한 물을 이용해 커터를 데워주면 깔끔하게 커팅할 수 있다.

성형 ❷

1. 데커레이션 나이프와 냉장 상태의 버터를 준비한다.

2. 데커레이션 나이프로 버터를 긁어 모양을 만든다.

FORMATION ❶

1. Prepare the butter at room temperature.

2. Spread the butter on a textured sheet in 2mm thickness, and harden in a fridge.

3. Remove the sheet when the butter becomes hard.

4. Cut in preferred shapes.
    * When using a cutter, warming it with warm water will help to cut neatly.

FORMATION ❷

1. Prepare a decoration knife and chilled butter.

2. Scrape the butter using the decoration knife to form its shape.

FRUIT & VEGETABLE CHIPS

과일 & 야채 칩

tools & ingredients

 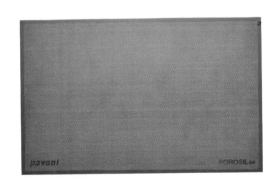 

TOOLS	슬라이서, 에어매트, 핀셋	slicer, air mat, tweezers
INGREDIENTS	30보메 시럽(물 1000g, 설탕 1350g), 과일	30 Baume syrup (1000g water, 1350g sugar), fruits

Process

1. 과일 또는 채소를 얇게 슬라이스한다.

2. 30보메 시럽에 담궜다 꺼내 에어매트 위에 올려준다.

3. 60℃ 오븐 또는 건조기에 넣어 충분히 건조시킨다.

1. Thinly slice fruits or vegetables.

2. Dip in 30 Baume syrup and place on an air mat.

3. Dry sufficiently in an oven at 60℃, or in a dehydrator.

IRON

열도장

tools

———

인두

iron

Process

1.   작업 전 인두를 충분히 달궈준다.

2.   비스퀴 표면을 가볍게 눌러 문양을 만든다.

1.   Heat the iron sufficiently before use.

2.   Press lightly on the surface of the biscuit to make imprints.

STARCH SHEET

전분 시트

tools & ingredients

TOOLS	펀칭 도구	design punchers
INGREDIENTS	전분 시트	starch sheets

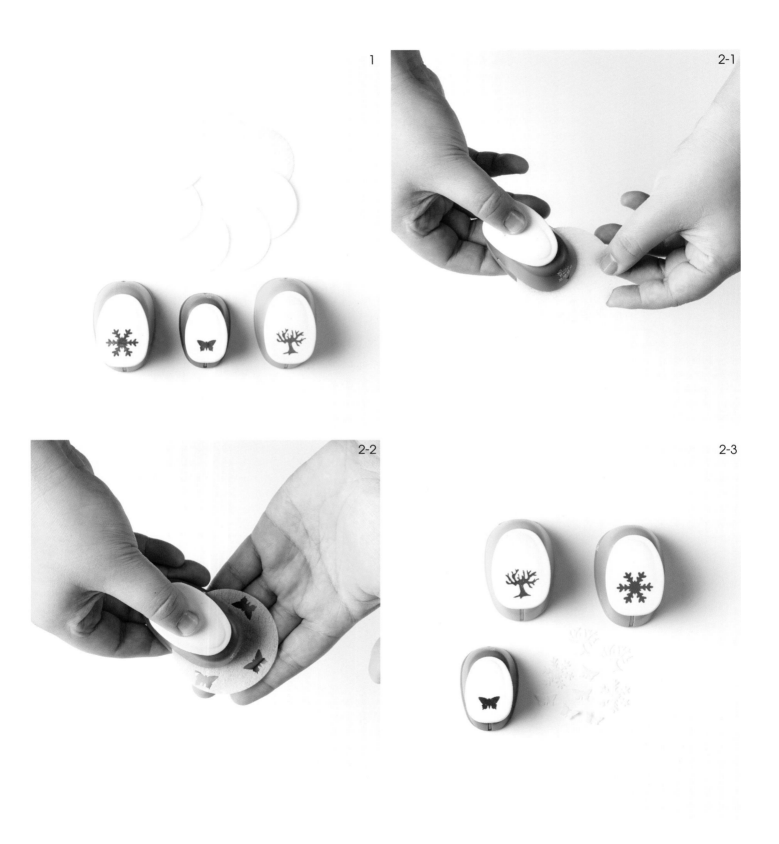

Etc

Process

1.   전분 시트와 펀칭 도구를 준비한다.

2.   펀칭 도구를 이용해 전분 시트를 커팅한다.

     * 전분 시트는 습기에 약하므로 즉시 서빙하는 디저트에 적합하다.

1.   Prepare starch sheets and design punchers.

2.   Cut the starch sheet with the puncher.

     * The starch sheets are vulnerable to humidity, so it is suitable for desserts serving immediately.

PART 07

NATURE

각 계절이 주는 자연의 선물을 놓치지 않으시길 바랍니다. 계절 허브와 꽃, 과일은 메뉴의 맛과 향을 부여하는 주재료로, 계절의 분위기를 전달하는 싱그러운 데커레이션으로 사용할 수 있습니다.

Please don't miss nature's gift from each season. Seasonal herbs, flowers, and fruits are the main ingredients that add flavors and aroma to the menu and can be used as fresh decoration which deliver the ambiance of the season.

HERBS AND FLOWERS

허브 & 꽃

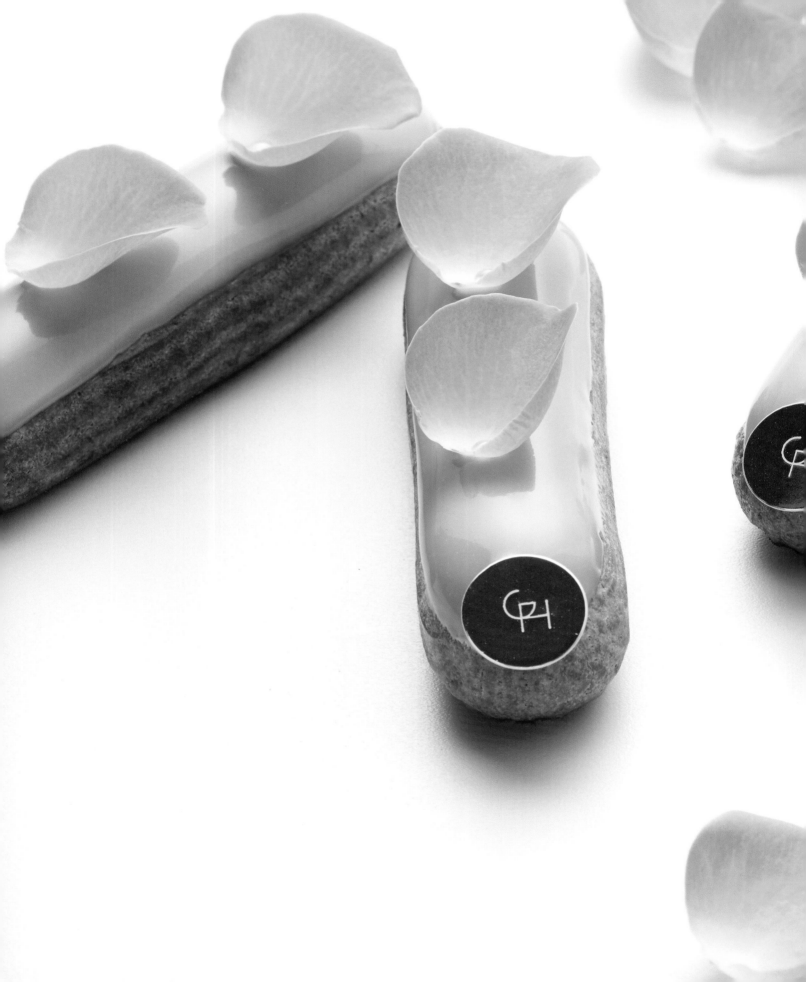

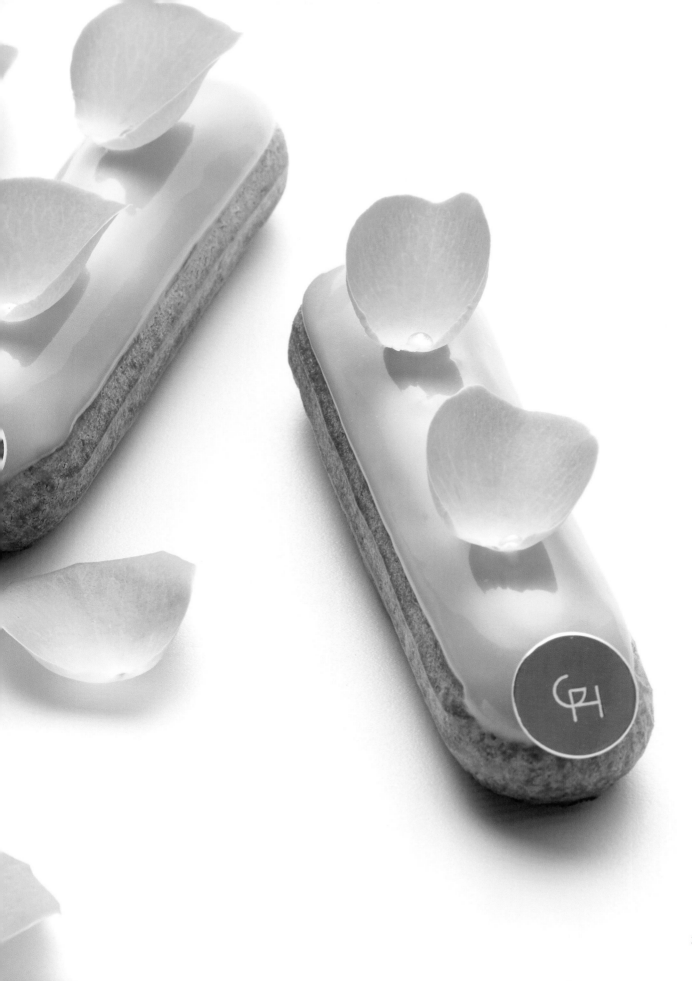

스위트 피
[SWEET PEA]

헬리오트로프
[HELIOTROPE]

콩꽃
[PEA FLOWER]

고수꽃
[CORIANDER FLOWER]

메리골드꽃
[MARIGOLD FLOWER]

노랑싸리꽃
[YELLOW SILKY BUSH CLOVER
(LESPEDEZA CUNEATA)]

레몬타임
[LEMON THYME]

카렌둘라
[CALENDULA]

채심꽃
[CHOY SUM FLOWER]

무꽃
[RADISH
FLOWER]

바질꽃
[BASIL FLOWER]

팬타스 [PENTAS]

수레국화
(CORNFLOWER)

금어초
(SNAPDRAGON)

휀넬꽃
(FENNEL FLOWER)

크레알썸
(WATERCRESS
FLOWER)

니겔라꽃
(NIGELLA FLOWER)

프렌치라벤더
(FRENCH LAVENDER)

구름패랭이꽃
(ALPINE PINK)

베고니아
(BEGONIA)

파인애플구아바
(PINEAPPLE GUAVA)

스토크
(STOKE FLOWER)

미니적시소
(MICRO RED SHISO)

초코민트
(CHOCOLATE MINT)

레드바질
(RED BASIL)

레드쏘렐
(RED SORREL)

자스민잎
(JASMINE LEAVES)

애플민트
(APPLEMINT (APPLE MINT))

기노메
(KINOME
LEAVES)

홀리바질
(HOLY BASIL)

레몬밤
(LEMON BALM)

바다휀넬잎
(SEA FENNEL)

레몬버베나
(LEMON VERBENA)

그린휀넬
(GREEN FENNEL)

CRYSTAL HERBS & FLOWERS

크리스탈 허브 & 꽃

tools & ingredients

TOOLS	붓	brush
INGREDIENTS	살균 흰자, 설탕	pasteurized egg whites, sugar

Process

1.　준비한 허브 또는 꽃에 살균 흰자를 얇게 발라준 후 설탕을 입혀준다.

1.　Brush a thin layer of pasteurized egg whites on the prepared herbs
　　or flowers, and coat with sugar.

2. 상온에서 약 12시간 정도 건조시킨 후 사용한다.

2. Dry for about 12 hours at an ambient temperature before use.

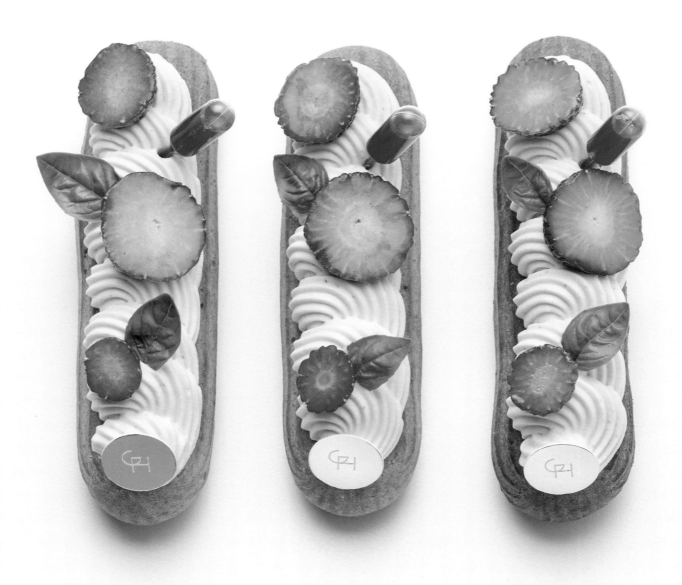

FRUIT SEGMENTS

과일 세그먼트

tools & ingredients

———

TOOLS	칼	knife
INGREDIENTS	과일	fruits

1-1 1-2

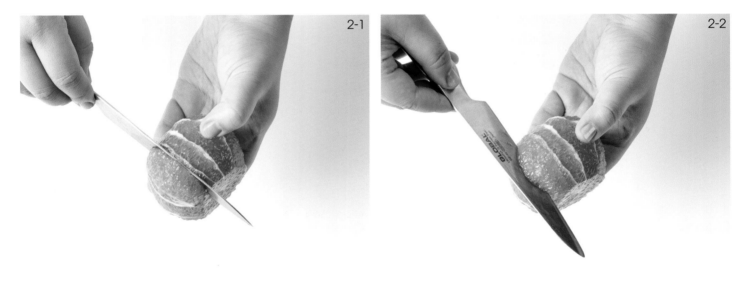

2-1 2-2

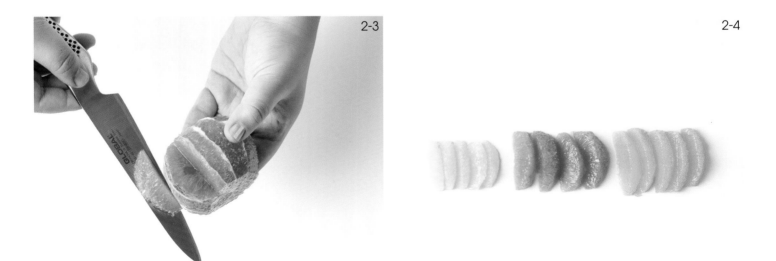

2-3 2-4

Process

1. 과일은 깨끗히 세척한 후 겉껍질을 제거한다.

2. 속껍질 사이사이에 칼집을 넣어 과육을 분리한다.

1. Wash the fruits thoroughly and remove the skin.

2. Cut between the wall and the pulp to separate the fruit segments.

GARUHARU

GARUHARU

GARUHARU

LEMON PEEL

레몬 껍질

tools & ingredients

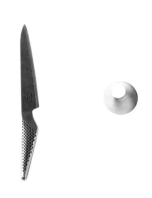

TOOLS	칼, 원형 커터	knife, round cutter
INGREDIENTS	30보메 시럽(물 1000g, 설탕 1350g)	30 Baume syrup (1000g water, 1350g sugar)

Process

1. 레몬을 깨끗히 세척한 후 껍질과 과육으로 분리한다.

   * 오렌지, 라임, 자몽 등을 선택할 수 있다.

2. 껍질의 하얀 부분을 최대한 제거한 후 원하는 모양으로 성형한다.

3. 끓는물에 한 번 데쳐 낸다.

4. 30보메 시럽에 치자 가루를 넣어 색을 낸 후 데친 레몬 껍질 장식물을 담궈준다.

   * 냉장 보관하며 사용한다.

1. Wash the lemon thoroughly and separate the skin and the pulp.

   * Any citrus fruits can be used- orange, lime, grapefruit, etc.

2. Remove the white rind as much as possible, and cut into the desired shapes.

3. Blanch once in boiling water.

4. Combine gardenia powder in 30 Baume syrup to color, and soak the blanched lemon peel decorations.

   * Store in fridge and use as needed.

COCONUT

코코넛

tools & ingredients

TOOLS	필러	peeler
INGREDIENTS	코코넛	coconut

Process

코코넛의 단단한 껍질을 제거하고 필러로 모양 내어 깎아 사용한다.

Remove the hard skin of the coconut, and shave with a peeler to use.

DECORATION
by
GARUHARU

외형적인 아름다움을 더해주되 메뉴가 가진 맛과의 연관성을 놓치지 않는 것.

심플하면서도 맛이나 텍스처에 방해를 주지 않는 것.

단지 장식이 아닌 맛과 향, 텍스처를 가지고 메뉴 안에서 하나의 구성이 되게 하는 것.

이것이 가루하루가 추구하는 데커레이션의 방향입니다.

Adding the external beauty but not missing the connection with the taste of the menu.

Simple but does not interfere with the taste nor texture.

Not solely as a decoration, but to make a single composition with taste, aroma, and texture within a menu.

This is the direction of the decoration GARUHARU pursues.

Team GARUHARU

가루하루의 시작부터 지금까지 새로운 시도와 도전에 늘 열정적으로 동참
해주는 가루하루 팀에게 고마운 마음을 전합니다.
성실하고 열정적인 재능 있는 동료들과 함께여서 새로운 시도를 주저하지
않고 무모했던 도전의 과정을 즐길 수 있었습니다.

I would like to express my gratitude to Team
GARUHARU for their passionate participation in
new attempts and challenges since the beginning of
GARUHARU. Team GARUHARU for their passionate
participation in new attempts and challenges.
With these talented teammates who are sincere and
enthusiastic, I was able to enjoy the process of reckless
challenges without hesitating to try new things.

GARUHARU MASTER BOOK SERIES 3

DECORATION by GARUHARU

First edition published	August 20, 2020
Fourth edition published	December 18, 2023

Author	Yun Eunyoung
Assistant	Kim Eunsol
Translated by	Kim Eunice
Publisher	Han Joonhee
Published by	iCox Inc.

Plan & Edit	Bak Yunseon
Design	Kim Bora
Photographs	Park Sungyoung
Stylist	Lee Hwayoung
Sales/Marketing	Kim Namkwon, Cho Yonghoon, Moon Seongbin
Management support	Kim Hyoseon, Lee Jungmin

Address	122, Jomaru-ro 385beon-gil, Bucheon-si, Gyeonggi-do, Republic of Korea
Website	www.icoxpublish.com
Instagram	@thetable_book
E-mail	thetable_book@naver.com
Phone	82-32-674-5685
Registration date	July 9, 2015
Registration number	386-251002015000034
ISBN	979-11-6426-126-0